Studies in Systems, Decision and Control

Volume 72

Series editor

Janusz Kacprzyk, Polish Academy of Sciences, Warsaw, Poland
e-mail: kacprzyk@ibspan.waw.pl

About this Series

The series "Studies in Systems, Decision and Control" (SSDC) covers both new developments and advances, as well as the state of the art, in the various areas of broadly perceived systems, decision making and control- quickly, up to date and with a high quality. The intent is to cover the theory, applications, and perspectives on the state of the art and future developments relevant to systems, decision making, control, complex processes and related areas, as embedded in the fields of engineering, computer science, physics, economics, social and life sciences, as well as the paradigms and methodologies behind them. The series contains monographs, textbooks, lecture notes and edited volumes in systems, decision making and control spanning the areas of Cyber-Physical Systems, Autonomous Systems, Sensor Networks, Control Systems, Energy Systems, Automotive Systems, Biological Systems, Vehicular Networking and Connected Vehicles, Aerospace Systems, Automation, Manufacturing, Smart Grids, Nonlinear Systems, Power Systems, Robotics, Social Systems, Economic Systems and other. Of particular value to both the contributors and the readership are the short publication timeframe and the world-wide distribution and exposure which enable both a wide and rapid dissemination of research output.

More information about this series at http://www.springer.com/series/13304

Chiang H. Ren

How Systems Form and How Systems Break

A Beginner's Guide for Studying the World

 Springer

Chiang H. Ren
Planned Systems International
Arlington, VA
USA

ISSN 2198-4182   ISSN 2198-4190   (electronic)
Studies in Systems, Decision and Control
ISBN 978-3-319-82964-7   ISBN 978-3-319-44030-9   (eBook)
DOI 10.1007/978-3-319-44030-9

Printed on acid-free paper

This Springer imprint is published by Springer Nature
The registered company is Springer International Publishing AG Switzerland

To Kelly

Contents

About the Author

 **Dr. Chiang H. Ren** is an entrepreneur, C-Level executive, and systems analysis expert who is currently the Chief Solutions Architect for Planned Systems International Inc. Prior to this appointment, he had been the chief technology officer for two other companies. In these positions and as a senior analyst for multiple United States government agencies, he has further published numerous peer-reviewed journal articles in operations research, disaster preparedness, information technology management, systems engineering, theoretical biology, and particle physics.

Dr. Ren holds a B.S.E. degree Magna Cum Laude in Mechanical Engineering and Applied Mechanics from the University of Pennsylvania, an S.M. degree in Aeronautics and Astronautics from the Massachusetts Institute of Technology, and a Ph.D. degree in Systems Analysis from the University of Bolton. Further he is a certified Six Sigma Black Belt, a member of the Tau Beta Pi engineering honor society, and an Associate Fellow of the American Institute of Aeronautics and Astronautics.

List of Figures

Chapter 1
The Mysterious Discipline

Abstract The introduction reviews the history of systems research and explains how the discipline has matured and divided into branches such as systems engineering, systems thinking, systems operations research, analysis of economic systems, theories for social and anthropological systems, modeling of biological systems, and management of organizational systems. This fragmentation of the discipline makes it difficult for a young student today or an interested novice without years of experience in a field that connects to systems studies to learn about systems analysis. There are many books written within each branch of systems research, but this book is unique and very much needed because it establishes a framework for studying systems that connects with all research branches but does not require the reader to have any prior backgrounds.

Friends, when I was a young engineer studying aerospace systems in graduate school, I greatly admired how the book *Six Easy Pieces*, by Nobel Laureate Richard Feynman [1], brought the wonders of physics to a more general audience. It is good to sometimes step back from the equations and look at the bigger picture. Then, as I entered the world of business, I greatly admired how the book *The 7 Habits of Highly Effective People*, by Stephen Covey [2], helped people focus on the real factors of success in life. It is good to sometimes stop working and think about what one is doing. Thus, after 25 years of studying systems and publishing systems research papers in fields such as disaster response, public administration, program management, national security, astrophysics, and theoretical biology, I find myself wondering why no one has published a book on the basic discipline of studying systems. The skill of being able to self-identify and explore behaviors and problems in our world from a systems perspective is so useful that it should be taught at the high school level.

Do not get me wrong. There are many fine books on systems engineering, systems thinking, systems operations research, analysis of economic systems, theories for social and anthropological systems, modeling of biological systems, management of organizational systems, and so forth. Further, there are books on specific processes and techniques derived from the study of systems, such as Lean

© Springer International Publishing Switzerland 2017
C.H. Ren, *How Systems Form and How Systems Break*, Studies in Systems, Decision and Control 72, DOI 10.1007/978-3-319-44030-9_1

Six Sigma, Total Quality Management, Balanced Scorecard, Fifth Discipline, and Agile Development. If I were a young student today or an interested novice without years of experience in a field that connects to systems studies, this great diversity of books that approaches the study of systems from different angles can be overwhelming. How do I begin to acquire systems analysis skills? And how do I move beyond the dictionary definition of a system being a group of parts that work together to yield total effects? Even so-called primers and introductory volumes are often oriented toward specific methodologies and philosophical perspectives.

The dictionary definition of a system indicates that we will find systems everywhere in our world and that all of nature can be considered a giant system. Some systems are made by man. Some systems are observable by man as clear constructs of nature. And some systems are more flexibly defined by man to help us better understand nature, society, and organizations. As a result, the study of systems is across many fields (transdisciplinary) and integrates methodologies from many fields (interdisciplinary). This strength of endeavor is perhaps also why systems studies is so fragmented and lacking a well-defined rudimentary core. I will further elaborate on this statement. But first, let me propose that, even for academic researchers and practitioners, it might be useful at times to step back from competing theories, contending schools of thoughts, set processes, and established tools to think about the basics. So, the search for the basics for those familiar and not so familiar with systems studies is the objective of this book.

The fragmentation of systems studies is tied to the fact that so many of us came to it from our own fields of study and bring to it our own biases in methodologies and research philosophies. I, for example, started with the design and engineering of well-bounded systems and later began to investigate the techniques for rapidly analyzing large military system of systems architectures in the course of providing technology and acquisition planning recommendations within the Pentagon during the latter years of the Cold War. Then, as I became involved in studying Information Warfare and exploring ranges of potential futures, yet another dimension of systems analysis opened up to me. This experience in the mid-1990s promoted a life-long research interest in complex system behaviors that can be projected through system models but cannot yet be validated because of a lack in supporting data. Sometimes, mechanisms for collecting the appropriate data have yet to be formulated, and, other times, the need to collect the appropriate data must be presented. I am sure that many others, such as biologists learning to build node and link diagrams as a part of the emerging field of systems biology and managers learning to build process flow diagrams as a part of business system reengineering, all have wonderful stories of how systems studies entered their lives. Further, I am sure that those who have majored in systems engineering and operations research will have a thousand stories of challenges, accomplishments, and collaborative experiences.

There is, however, a much bigger story of systems studies that extends back to the establishment of the scientific methodology by Johannes Kepler in 1602, [3] and it is worthwhile to summarize this story to help place all our experiences and the objective of this book in context. Kepler's Laws of Planetary Motion is an observationally and analytically established model for planetary systems. Since the days of Kepler, scientists have been deductively breaking apart all aspects of natural systems into measureable and relatable pieces to support hypotheses, theories, and validated facts. For complex natural systems such as living organisms, the efforts to identify their component parts intensified with the discovery of the cell in 1676 and cell structures in the 1800s [4]. The philosophy that a system is no more than its identifiable component parts, scientific reductionism, in turn, became very popular in natural science communities [5]. As scientific instrumentation advanced in the twentieth and twenty-first centuries to identify all component parts, the philosophy of positivism, which states that knowledge should only be based on what can be measured and mathematically/logically explained, also became popular in natural science communities [6]. These philosophies continue to influence those with prior scientific training in their study of systems.

Systems studies followed another path with the industrial age, as man created ever more sophisticated systems to serve society. Inventions, such as those by Thomas Edison starting in 1869, were achieved through inspiration, creativity, and inductive thinking [7]. The figuring out of how parts fit together and the designing of parts for fitting together into systems have been the focal points for the engineering fields. This endeavor has intensified with the miniaturization of electronic devices, the start of the computer age, and the growth of the World Wide Web. To study systems that must work together to form greater systems, the US Department of Defense and others have invested substantial resources since WWII in operations research (how systems perform in real environments), logistics (how systems are supported during operations), lifecycle management (how systems are built, deployed, and retired), and war gaming (how systems specifically compete with other systems in symmetric and asymmetric ways). Recognizing that modern man-made systems must often integrate mechanical, electronic, computer, and communication subsystems as well as take into account the capabilities and limitations of the users, many universities and institutions have established systems engineering departments and divisions. The term "systems engineering" traces back to Bell Telephone Laboratories in the 1940s, [8] and engineering endeavors have since focused on design, modeling and simulation, optimization, control, and reliability. With the advancement of computer tools over the past decades, all these endeavors have matured into specialized fields, and some of the modeling techniques have been adapted to study biological systems.

As the industrial age shifted the structure and tempo of societies, the study of systems followed a third path into the social sciences. Herbert Spencer popularized the philosophy of functionalism, which argued that society should be viewed as a complex system with mutually supporting parts [9]. He also introduced the biological theory of natural selection into social dynamics. As society has economic, political, military, and cultural components, each of the connected academic fields has incorporated systems thinking and systems modeling into their studies. For example, Karl Marx, in 1867, presented one of the earliest theories on social system failure by arguing that economic inequalities will cause internal tensions that lead to social collapse [10]. Von Neumann, in 1944 [11], mathematically modeled the interactions across political systems and systems driven by individual actors based on rational decision-making by all sides. The interactions gave rise to Game Theory, which was advanced by many scholars and later applied also to biology. Yet Karl Ludwig Von Bertalanffy and others in the 1930s made perhaps the most important advancement in systems thinking for the social sciences through the argument that social systems are too complex to be studied by pure scientific reductionism or engineering-based mechanistic models. Instead, the resulting General Systems Theory argued for the study of social systems to be more focused on holism and organic behaviors [12].

General Systems Theory launched the realization that systems involving inter-acting human actors cannot be tightly bounded or easily quantified despite the endeavors of man to create structured organizations. Like other organic systems, the complexity is often reflected in self-organizing, self-adapting, and even self-proliferating characteristics. However, modeling such characteristics can be more challenging than systems in nature because we do not always have an objective system state or reference frame for how the human system should per-form. After advancing operations research in the 1950s, Churchman [13] would declare such systems are "wicked problems," and Ackoff [14] would call such problems "messes". To study these systems, Checkland in the 1980s [15] formu-lated the Soft Systems Methodology (SSM), which recognizes that our actions to measure a system affects the system and that there are no perfect models of systems. Instead, SSM advocates a recursive learning approach for systems understanding starting with an initially imperfect conceptual model. This methodology is aligned with the philosophy of action research and challenges the idea that man can engineer rigid systems and organizations that have complete mastery of interactions within their environment [16]. Contrary to the objectives of design research, there may always be some hidden consequences, latent patterns, and/or unforeseen forces because the true nature of all real-world systems is unbounded.

I have mentioned deductive and inductive methodologies in systems studies; thus, SSM should be considered a more explorative methodology. However, there are other ways to explore complex adaptive systems as first defined by the Santa Fe Institute [17]. If we are not certain about whether a bunch of parts even constitutes a system or whether many interacting systems will lead to unrealized effects, modern computers now enable us to simulate such behaviors through agent-based models. The philosophy of agent-based modeling is the belief that even simple interactions between agents (computer models representing people, organizations, things, and the environment) lead to highly complex outcomes over time. If we want to study macro behaviors in an extremely large and complex system, modern computers now enable us to simulate dynamics at an abstract level using models built based upon the principles of system dynamics as established by Forrester in the 1960s [18]. One type of abstraction is a way to model the whole world based on inter-regional and transnational division of labor through the World Systems Theory of the 1970s, to be discussed later. However, there are many other theories on how to abstractly model geopolitical and transnational behaviors.

In the United States, system dynamics has greatly influenced the social sciences, economic theories are being extended to biological systems, and researchers are still trying to validate the results of agent-based models. However, Soft Systems Methodology has historically remained unpopular and is left largely to the endeavors of European researchers. We can speculate that the United States has invested tremendously in the design of physical and organizational systems over the years, and control or illusions of control, depending on your perspective, has taken priority over systems understanding in some cases. Certainly this appears true at the organizational management level where many books have been written to help practitioners control organizations for optimization and transformation. Simplistically, some models are more system metrics focused, such as Total Quality Management and Balanced Scorecard [19, 20], some models are more systems process focused, such as Lean Six Sigma [21] and some models are more systems integration focused, such as The Fifth Discipline as established by Senge in 1990s [22]. There are overlaps between these models, and all the models seek to transform organizations. However, the consideration of organic behaviors in the organization varies, and the control points, as a result, vary.

At this point, I will apologize for not doing justice to any of the system study paths and methodologies presented. However, these paths and methodologies will reappear again as we explore the basics in studying systems. All I wish to show for now is the reality that systems studies lack a single coherent core, and that, despite efforts to apply methodologies and techniques across disciplines, the dichotomy between the paths has caused contention and mutual misunderstanding. Practitioners and researchers along different paths of systems studies are indoctrinated into communities, and problems in communities are still falling through the cracks because of philosophical limitations. To the rest of world not familiar with systems studies, it must truly appear like a mysterious discipline. There is so much promise for problem resolution and so much ambition in the scope of problems being tackled. Yet, I will argue that seldom has system study approaches and outcomes been explained clearly and concisely to young students and senior decision-makers. One of the most familiar system diagrams in the news years back is that of a messy chart trying to show the interrelationships between factors affecting stability in Afghanistan. Instead, the chart convinced the general public that the Pentagon had missed the big picture [23].

I am not sure that everyone conducting and applying systems research can ever agree on philosophies, methodologies, and theories. But I do know that I am not the one who can bring about agreement. Sometimes a little disagreement is healthy for the advancement of knowledge, as long as each side is willing to consider the arguments of the other. Other times much potential is lost. My interest is to introduce the wonders of this mysterious discipline to the outside world at a basic level where there are no major disagreements. As hopeful novices, let us now explore how systems form and how systems break. Then, you the reader can decide to what degree you want see the world through the perspective of systems analysis and to what depth you wish to learn about systems analysis techniques.

References

1. Feynman RP (1963) Six easy pieces. Addison-Wesley Publishing Company, New York
2. Covey SR (1989) The 7 habits of highly effective people. Simon & Schuster, New York
3. Kepler J (1619) The harmony of the world. CreateSpace Independent Publishing Platform
4. Hooke R (1665) Micrographia: or some physiological descriptions of minute bodies made by magnifying glasses, with observations and inquiries thereupon. Courier Dover Publications, Mineola
5. Jones RH (2013) Analysis & the fullness of reality: an introduction to reductionism & emergence. Jackson Square Books, Oxford
6. Ayer AJ (ed) (1966) Logical positivism (the library of philosophical movements). Free Press, New York
7. Dyer FL, Martin TC (2012) Edison, his life and inventions. A Public Domain Book
8. Fagen MD (1978) A history of engineering in the Bell system. Bell Telephone Lab, Murray Hill
9. Francis M (2007) Herbert Spencer and the invention of modern life. Cornell University Press, Ithaca
10. Wolfgang R (2009) Karl Marx. Morgan Reynolds, Greensboro
11. Von Neumann J (1944) Theory of games and economic behavior. Princeton University Press, Princeton
12. Von Bertalanffy L (1969) General Systems Theory: foundations, development, applications (revised edition). George Braziller Inc., New York
13. Churchman CW (1967) Wicked problems. Manag Sci 4(14):141–142
14. Ackoff R (1974) Redesigning the future. Wiley, New York
15. Checkland P (1981) Systems thinking systems practice. Wiley, Chichester
16. Foster M (1972) An introduction to the theory and practice of action research in work organizations. Hum Relat 25(b):529–556
17. Waldrop MM (1993) Complexity: the emerging science at the edge of order and chaos. Simon & Schuster, New York
18. Forrester JW (1975) Counter intuitive behavior of social systems. Collected papers of Jay W. Forrester. Wright-Allen Press, Cambridge
19. Martínez-Lorente AR, Dewhurst F, Dale BG (1998) Total Quality Management: origins and evolution of the term. TQM Mag 10(5):378–386
20. Kaplan RS, Norton DP (1996) The Balanced Scorecard. Harvard Business School Press, Boston
21. Pyzdek T, Keller P (2010) The Six Sigma handbook, 3rd edn. The McGraw-Hill Companies, New York
22. Senge P (1990) The Fifth Discipline: the action and practice of the learning organization. Doubleday, New York
23. Bumiller E (2010) We have met the enemy and he is PowerPoint. New York Times, April 26

Chapter 2
The Characteristics of Systems Formation

Abstract This chapter establishes the basic concept of what constitute systems and defines characteristics associated with the concept. The characteristics are broad enough to apply to all types of systems and have associated metrics that can define specific system components, structures, and behaviors. Many examples based on natural systems, human organizational systems, and man-made systems are provided in each section to explain how system metrics are to be applied. Methodologies for studying systems are further introduced in the context of applying metrics to specific types of systems. Through the established conceptual framework, we further explain how hidden systems can be discovered, logical divisions between systems can be determined, systems can be designed based on total dimensionality, and the behaviors of systems can be explored.

Our world is filled with systems and activities that can be defined as systems. Therefore, a student studying systems formation might be tempted to just jump into case studies upon case studies. The challenge with this approach of going from the specific to the general is that one might never get the case studies to converge upon a common understanding and one may never be certain that the right scope of case studies have been used to achieve common understanding. Studying real world systems, even at a fundamental level, further requires subject matter skills. The division between subject matter experts then enforces the fragmentation of the discipline.

Our study of systems formation will, therefore, start with the basic concept of what constitute systems and the definition of characteristics associated with the concept. These characteristics will perhaps be obvious to some by first introduction. Yet, if all systems are bound by these characteristics, then we can study systems formation by going from the general to the specific. I believe that these general characteristics will help us discover hidden systems, determine logical divisions between systems, design systems based on total dimensionality, and explore the behaviors of systems. And, we need to first understand how systems form before we can study how systems break. People who are studying specific failure modes might

© Springer International Publishing Switzerland 2017
C.H. Ren, *How Systems Form and How Systems Break*, Studies in Systems, Decision and Control 72, DOI 10.1007/978-3-319-44030-9_2

argue with me about the last statement, but the statement might make more sense after I explain my definition for formation.

I fully understand that many systems in nature are so complex and so old in origin that their paths of formation will continue to elude us. However, nature and even our own physical bodies teach us that whatever is not forming or growing is often in the process of failing and dying. As soon as our bodies reach adulthood, the process of aging begins. As soon we build a machine, the process of wear and breakdown begins. Breakdown can be controlled and delayed through maintenance, but absolute steady-state is a rare thing. So the study of system formation is the study of the system across its life of changes and transformations to the point where breakdown is unavoidable. Sometimes, failures occur in the process of formation, and other times failures occur after formation has stopped. Either way, to fully understand failures, we need to know not necessarily the beginning of formation but most definitely the end state of formation and formation activities. That end state, even when cut short, is the reference frame to which system breakdown can be measured.

If a system is a group of parts working together as a whole according to definition, then an understanding of system formation must involve the study of:

- The dynamics of the parts and the whole
- The associations between the parts to make the whole
- The structure of the whole based on the parts and associations
- The boundaries of the whole or the boundlessness of the whole
- The interactions between the system and the environment with other systems
- The qualities of the system as a whole
- The integration of systems to form greater systems.

These can be considered the top-level characteristics of systems formation, and the many paths and methodologies of systems studies can be placed in the decomposition of characteristics. These characteristics also affirm that systems studies is a discipline that cuts across other disciplines and integrates disciplines. As researchers have long realized, systems with common characteristics in nature and society often exhibit similar behaviors that enable comparative analysis. Systems with unique capabilities in nature and society might further inspire the design of man-made systems. And man-made systems often integrate with social and nature systems in complex ways that have potentially unforeseen secondary effects.

As a result, our journey into the basics of systems formation will be an examination of system characteristics and interrelationships between characteristics. If you approach all the problems, opportunities, and behaviors of this world through the lens of these characteristics, I guarantee you that the world will never appear the same again. If you partake of other fields of study through the lens of these characteristics, then each field will not appear so distant and so alien to your understanding. The patterns of system behaviors repeat themselves over and over again, and the causes of system failures, which we will explore in Chap. 3, are seen everywhere that we find systems.

2.1 Dynamics: Moving System Parts

Any discussion of systems should probably start with the term "dynamics" because there cannot be a system without change. A bunch of parts connected together in an unchanging way is merely an object. The object can be incredibly complex. However, if there is no work being done and no changes occurring, then the object is a display piece. On the other hand, a combination of very simple moving parts, such as a wheel that grinds wheat being turned by water flowing down stream, forms a system, and the activities of the system are termed system dynamics. Systems dynamics is the dynamics of the parts and the dynamics of the whole system. In very simple systems, the dynamics of the parts is easily translated into the dynamics of the whole systems. In very complex systems with many parts, complex parts, unknown parts, and/or unknown parts relations, the study of the system becomes a dedicated discipline. Before we go too far down the road of complex systems, first let us start with an understanding of the basic dynamics for system parts.

A part that belongs to or could belong to a system is generally described through four types of dynamic characteristics as shown in Fig. 2.1. As the part can be a material component, software module, human actor, biological entity, information element, or a subsystem composed of any combination of the other part types, we must start with a very broad understanding of dynamic characteristics and then advance our understanding toward specifics.

In the first type of dynamic characteristics, the part will have an orientation relative to a reference frame that is based on the system or the system's operating environment. For machines, one orientation would be how a part fits with other

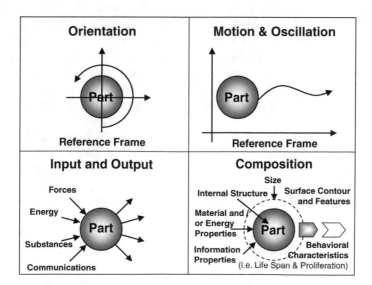

Fig. 2.1 Four types of dynamic characteristics

parts, and the orientation could be relatively fixed in the design reference frame. However, as the machine moves in an environmental reference, the orientation of the part will change relative to the environment as well as the forces and material interactions within the environment. For software, the orientation could be the position of a group of codes relative to other codes and code interfaces. Software parts must reside in physical computer parts, but the management of software through hardware and platform technologies in modern network-based cloud computing systems does not have to follow a one-to-one relationship. For human organizations, the orientation could be the political leaning of a special interest group, the procedural guidance for a team, the needs of different users, etc. For systems of pure information, the orientation could be how a bundle of information is positioned against an agenda such as a marketing campaign with many bundles of information working together. The key point about orientation is that it might be a governing factor to how parts will work together and how the working relationships can change.

The second type of dynamic characteristics is motion and oscillation as a specific type of motion. Once again, motion needs to be measured against a reference frame, and a part can have different kinds of motion relative to the system and to the operational environment. In the physical world, motion could be linear or rotational. Linear motion merely means that a line vector describes the motion, but the path or pattern of motion could follow a complex trajectory. If a motion continuously reverses and repeats itself, then the part is in oscillation. A part can move linearly alone one directional axis and oscillate along another. Also, a part can oscillate in place back and forth or oscillate about a rotational axis. And, rapid back and forth motion can be described as vibrations. For nonphysical parts, motion is essentially a statement of change for the whole part relative to a nonphysical reference frame. For example, an encapsulated malware is in motion across the World Wide Web until it lashes onto a host software application and causes harm. Humans in society or an organization are said to be in motion if they change locations or if they change their group alignments. As I will discuss later in studying system structures, some systems and structural configurations can tolerate the relative motion of their internal parts more than others. Both internal motion and motion tied to the whole system might affect the input and output characteristics of a part and the composition of a part.

Accordingly, the third type of dynamic characteristics is input and output for different parts. If a part has the structure of a subsystem, then how that subsystem receives inputs and transmits outputs to other parts, systems, and the environment is fairly complex. Regardless, all manner of simple and complex inputs and outputs can be further categorized as forces, energy, substances, and communications. This breakdown is in favor of physical parts, as they can receive and transmit all four kinds of input and output. For information technology systems, the inputs and outputs are limited to energy and communications. However, the communications can be further subdivided into data transmission, software uploads and downloads, protocol exchanges, and status updates. For human systems, inputs and outputs could represent ownership. Products can be given to a human recipient. The human recipient can pass the products to others. And the human recipient can create or modify his/her own products to pass on as output. In fact, a part taking inputs and using them to create outputs is one of the most common component functions in a system.

The fourth type of dynamic characteristic is the composition of a part. At the most fundamental level, a part should have a size and surface contour with features that are relative to the reference frame for that part. If the part is a physical component, then the size can be from the atomic level to the planetary level because the atom is a system, and the stars and galaxies are all systems. The physical features could be receptors that promote integration with other parts or systems, textures that affect contact interactions, and gates that control the inputs and outputs. For software parts, the size could be number of lines of code, and the surface could simply be the code boundaries and interfaces. For parts in human systems, the size could be the number of people in a component group and the surface could be the positions of the people. Finally, an information part could be sized by the quantity of information and the accessibility of the information. Inside each kind of part, there should be an internal structure that could be very complex. Physical structures can have material and energy properties, information properties, and behavioral characteristics. Other structures might only have information properties and behavioral characteristics. The information properties of software parts might be very complex, and the behavioral characteristics of organic and human parts in systems might be even more complex. This complexity sometimes includes how parts can self-proliferate and how parts will age and break down overtime. The sources of complexity lead us to the next step of exploring how to study the dynamics of parts and systems.

As all systems have dynamic characteristics, measuring and studying the macro-dynamics of the total system is a way to identify and understand the system parts. Then, measuring and studying the dynamics of system parts is a step toward understanding the formation process of the system. The measurement of the whole and the pieces can be an interactive process that steadily incorporates the other characteristics of formation as the understanding of the system begins to manifest. However, the endeavors of measurement bring us into the positivism versus soft systems thinking debate.

I will at this point declare that I do not strongly embrace the positivism philosophy like so many of today's scientists. This is because I do not accept that today's instruments and methods can always measure all the parts and part characteristics in real-world systems. Further, I believe that, despite the lack of data, systems studies might still help us press forward with discovering new methods of measurement, new system needs, and even new parameters that have been ignored by other researchers. Instead of building systems thinking around the data, I, like many others, prefer to build systems thinking around the actual problems and dynamics observed in the real world. In this manner, I agree with soft systems thinking in that real-world systems can never be perfectly measured because a perfect set of measurements means that we will have built another model of the real world. To elaborate, every measurement of change that we take with modern instruments is still at intervals across specific parameters. Movies are at a frames-per-second rate that is much faster than the eye and brain can perceive. Digital images breakdown data intervals into pixels per square inch. And computer databases must record data in distinct increments.

The two big shifts in the measurement of systems in modern time are: (1) a dramatic increase in measurement capabilities across many scientific fields, and (2) a dramatic increase in data storage plus processing capability with high capacity blade servers, fiber optic networks, and cloud computing distribution platforms. The most dramatic advances in measurement are perhaps in the biological sciences with the conduct of the Human Genome Project from 1990 to 2003, the identification of countless proteins/enzymes that regulate cell activities, and the discovery of many drug combinations that affect biological processes. However, the details of our universe gathered by the Hubble Space Telescope and other space probes are also impressive advances. The most dramatic advances in database usage are perhaps in the social media business area where the buying patterns, viewing habits, demographics, and preferences of millions of online users can all be recorded as terabytes (1000 GB) of data. However, these databases will soon be rivaled by databases with the electronic health records of billions of people. To place a terabyte in perspective, an IBM PC in 1982 has a 5-MB hard drive. This means that one of today's 4 or 6 TB drives, which only cost a few hundred dollars, will have the data storage capacity of 1 million 1982 IBM PCs.

The net result of this explosion in data collection and storage is that the world now has and may continue to have more data than system models to understand the data. If we believe that the world is composed of systems within systems, then all data in theory has a systems connection. Achieving that connection is perhaps the biggest challenge for systems studies in the future. However, data can be deceptive because immense quantities of data do not mean that the data sets are complete. Hidden patterns in the dynamic characteristics of parts can exist between measurement intervals. Some dynamic characteristics may still not be measured. Some parts may still not be detected in measurements. And some system formations may not be identifiable even with tons of existing data. For example, with all the research attention devoted to capturing the DNA as the map for organic growth, operations, and senescence, I have instead wondered who is doing research on the reference frame for the DNA map [1]. How do cells in the body grow and specialize into shapes and functions using the DNA map? No matter how well we measure the map, the system understanding is incomplete without the mechanism for the reference frame. The search for missing information, undetected parts, and unidentified systems will require an integrated understanding of all the characteristics in systems formation. Therefore, at this point, let us first explore what to do with all the data at hand.

For data sets that are well structured in that the primary information has clear fields of associated information, computers have been quite successful at storing and using such data through relational databases. Relational databases use table structures to capture data and correlate data fields through a relational index. A spreadsheet is an example of a relational database. As shown in Fig. 2.2, the first

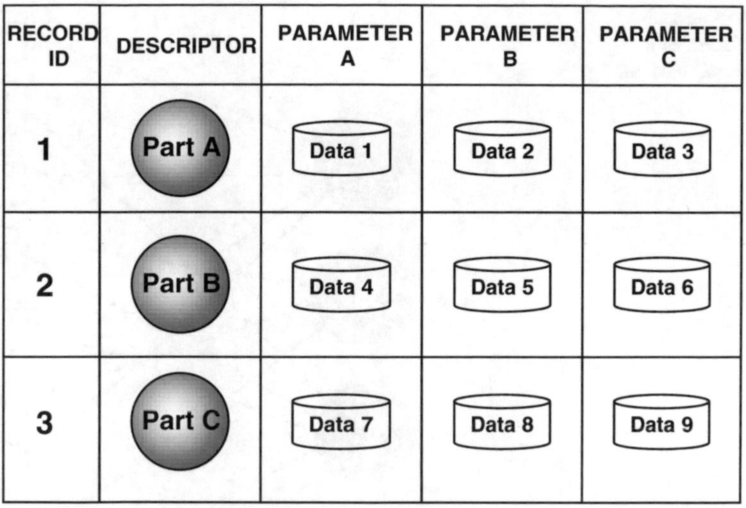

RECORD ID	DESCRIPTOR	PARAMETER A	PARAMETER B	PARAMETER C
1	Part A	Data 1	Data 2	Data 3
2	Part B	Data 4	Data 5	Data 6
3	Part C	Data 7	Data 8	Data 9

Fig. 2.2 Notional representation of a simple relational database

column of record IDs connects the elements/parts being described with the fields of descriptors and associated information. These databases can be quite large, as long as the relationship structures can be maintained, but there will eventually be scaling problems (perhaps at the terabyte level), as the size of the database cannot be easily handled by current server technology.

In response to large data sets without well-defined relational structures and with the need to leverage distributed cloud computing capabilities, technologies for nonrelational databases have advanced, led by Google and other leaders such as Apache. Essentially, nonrelational databases, as shown in Fig. 2.3, try to encapsulate data and parse data across a terrain. The data can be managed and controlled at the cell level with even security and access to the data controlled at the cell level. With this parsing, packets of data can be dynamically associated with one another in a complex manner based on incremental and iterative advances in understanding the data. The first step in advancing our understanding of the data is data mining. So in this first section on part dynamics, we will review data mining techniques and leave the many analytical techniques that are applicable to mined data for later sections.

Almost everyone today who has been on the Internet has conducted data mining activities. The most popular mining endeavor is the Google search based on key words and phrases. What the user gets in data mining are hopefully pieces of information from vast quantities of data that shed light on the user's problem and research interests. It is easy to understand the concept of a key word search, but there are other more advanced searches into the vast networks of data. I will review some of these advanced techniques below, and many of these techniques will require specialized search tools and inference engines that connect search activities with rule sets.

Fig. 2.3 Notional
representation of a
nonrelational database

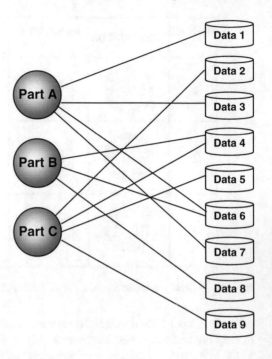

2.1.1   Data Mining by Deductive Decision Tree

In this technique, as shown in Fig. 2.4, a search engine is given a hierarchical set of rules, which is automatically applied to search results. With each level of the search, the results are automatically assessed, and the rules tell the search engine which branches to follow in the next level of search. This multistep search capability produces incremental results that are presentable in a tree structure to promote data relationship understanding.

This technique is quite useful in rapidly searching for parts and part characteristics that are associated with an evolving distributed system in a complex environment [2]. For example, this search can automatically map out how a disease system is spreading across a society. Also, this technique is quite useful in tracking down sources of errors in complex multistage organization processes.

2.1.2   Data Mining by Agile Characterization

In this technique, as shown in Fig. 2.5, a search engine collects data broadly and dynamically organizes the data into summary groups, such as groups based on data ranges, for presentation [3]. The purpose of the grouping is to enable rapid comparisons of contrasting data between groups and to adjust group boundaries to

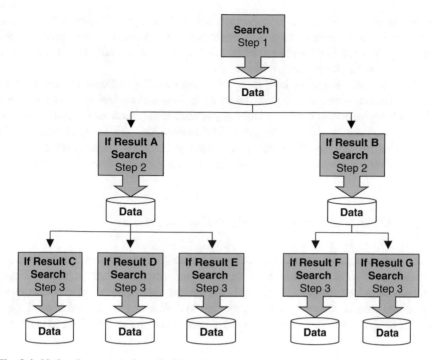

Fig. 2.4 Notional representation of mining tree

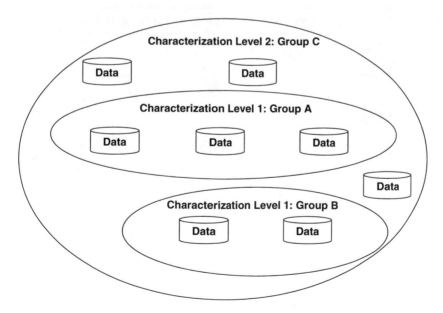

Fig. 2.5 Notional representation of characterized groups

better characterize data for follow-on searches. This process of characterization and recharacterization might require sufficiently generalized definitions of groups in the beginning, but the iterative searches that increasingly place data in more accurate groups can yield precise descriptive results.

This technique is quite useful in figuring out which distributed system, such as military forces, owns which parts as systems interact/conflict with one another. Also, this technique is useful in isolating system parts, such as biological agents, from an environment of similar parts. The refined definitions of groups can be further used to describe the associated system at a macro-dynamic level, and the process of grouping can be used to design or form systems from raw material.

2.1.3 Data Mining by Complex Classifications

In this technique, as shown in Fig. 2.6, a search engine identifies properties that are common across all or portions of the data and interrelationships between data elements based on these properties [4]. The initial identification process can use a correlation matrix. Once there are properties to link data elements, these links can be used to determine parts that belong to a system and the associations between the parts.

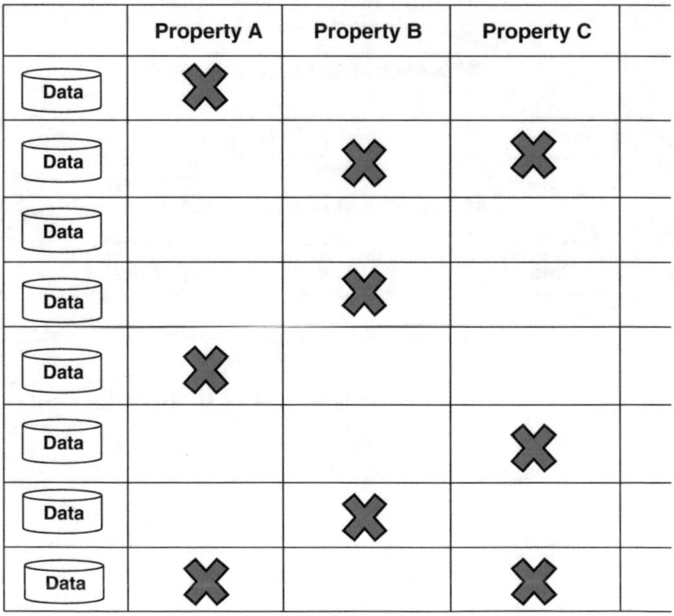

	Property A	Property B	Property C	
Data	✖			
Data		✖	✖	
Data				
Data		✖		
Data	✖			
Data			✖	
Data		✖		
Data	✖		✖	

Fig. 2.6 Notional representation of complex classifications

This technique is quite useful in filtering data elements, such as properties of people in society, for behavioral patterns that link select elements to systems, such as secret organizations. Also, this technique is useful in separating properties/effects that belong to parts in a system from other related properties/effects from the environment.

2.1.4   Data Mining by Regression Analysis

In this technique, as shown in Fig. 2.7, a mathematical best fit line or curve fitting tool is used to discover how to extend the known patterns in data into regions of

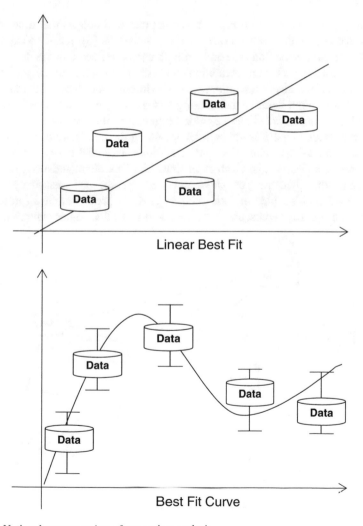

Fig. 2.7 Notional representation of regression analysis

unknown data [5]. Linear regression can project the nature of data in regions beyond current measurement capability. Alternatively, curves can show us ranges of potential data.

This technique is quite useful in guiding researchers toward areas of missing system dynamics information, such as output qualities when inputs are increasing beyond current measurements. Also, this technique is useful in formulating/ projecting the existence of additional parts for systems with the understanding that such parts are pending future verification.

2.1.5   Data Mining by Inductive Data Association

In this technique, as shown in Fig. 2.8, a computer tool creates real-time node and link constructs in data based on discovered associations [6]. As to be explained in the next section, association type and strength can be reflected in the definition and distance of linkages. This representation can further be used to identify spatial gaps in data and future collection requirements. The initial inductive model might not be accurate, but through iterative data mining based on the model, the study of system parts and the whole system can be folded together in the data mining process.

This technique is quite useful at quickly linking the behaviors of the parts to the dynamics of the total system. Also, projected links are useful in finding data as well as hidden system parts. The changes in links and link characteristics will provide insight into the dynamics of parts and the system, and massively complex point-to-point relationships in data, such as those in protein studies (proteomics), might be more easily represented by nodes and links than other capture methods.

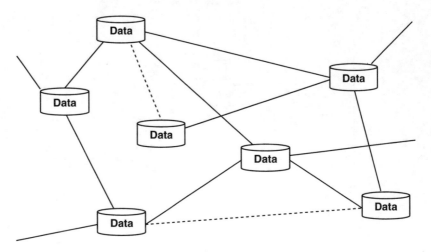

Fig. 2.8 Notional representation of inductive data association

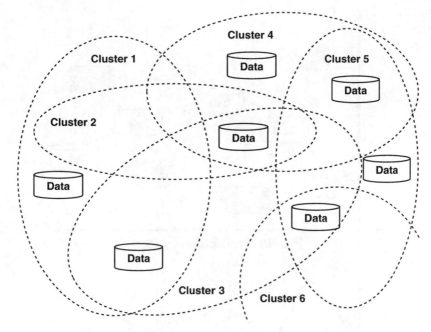

Fig. 2.9 Notional representation of cluster analysis

2.1.6 Data Mining by Clustering Analysis

In this technique, as shown in Fig. 2.9, the search engine conducts artificial grouping and regrouping of data to discover metadata sets where knowledge discovery is better achieved [7]. The meaning of a cluster is often understood after analysis whereas the meaning in data classification is more connected with the classification process.

This technique is quite useful at studying a mass of data, such as in information-driven systems, with no clear interrelations and delineations. At the beginning of the data collection processes, clusters can be flexibly assigned and overlapping. Then as data changes, the clusters can be refined to more accurately reveal system content and system dynamics understanding.

2.1.7 Data Mining by Baseline Pattern Searches

In this technique, as shown in Fig. 2.10, the search engine looks for entire patterns, groups, and states in data based upon traceable paths and/or baseline reference frames [8]. These entities may sometimes be obscured by other data elements

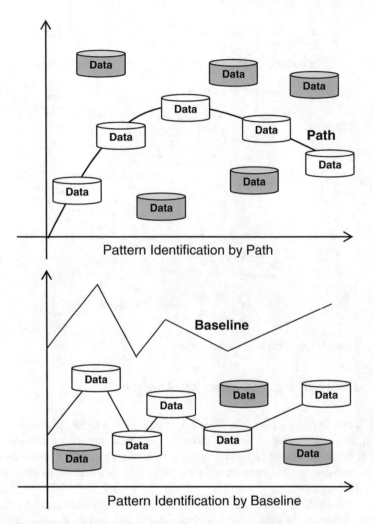

Fig. 2.10 Notional representation of pattern searches

intermixed into the patterns and groups. Therefore, a path or baseline is used very much like a filter to discover behaviors and relationships within apparent chaos.

This technique is quite useful in comparative data analysis, such as finding similar patterns of disease propagation in other cities when there is a baseline pattern from the originating cities. Also, this technique is useful in finding new, perhaps hidden, patterns by trying out a variety of nonrandom paths as filters. For example, admits the individual activities of people in a city, unique patterns of behaviors, such as specific person-to-person interactions or movements from location to location, can be discovered to indicate a coordinated terrorist plot.

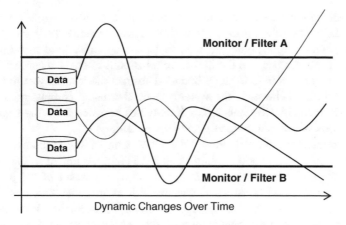

Fig. 2.11 Notional representation of state change and deviation filters

2.1.8 Data Mining by State Change and Deviation Filters

In this technique, as shown in Fig. 2.11, data filters are continuously applied to monitor for when dynamic patterns have exceeded specified ranges [9]. If the data is connected with system parts, then the dynamic characteristics of the parts, as described above, can be used as a basis for determining what states to monitor.

This technique is quite useful in understanding peak behaviors in defined system parts undergoing volatile periods of changes, such as worker dynamics in an organization hit by a business crisis. For example, who needs counseling support and who needs to be released can be assessed by behavioral filters. Also, this technique is useful in finding parts that are acceptable in a system, such as mechanical testing of manufactured components for performance within designed limits.

The above techniques for data mining are naturally mathematically involved during implementation, and many complex algorithms as well as computer codes have been developed in the exploding field of "Big Data" analytics. However, it is important for us to not lose sight of the fact that the human mind, which processes data in a nonlinear manner, is still superior to the computer's linear processing in some ways despite the computer's overwhelming speed, capacity, and accuracy. Therefore, I introduced the above concepts not merely to be a beginner's tutorial but also to be a stimulus for people closest to data to see pass the obvious for insights based on thinking about how to look and what to look for. To elaborate, the computer sees data as discrete elements and must work through data from one piece to the next. If the computer draws a curve through points, it goes from point A to B to C. In contrast, the human mind sees data as a whole as well as in discrete elements. Therefore, when we draw a curve through points, we can, if trained and focused, see how the curve fits simultaneously at all points. At times, we can still present a better fit solution in a faster time frame, particularly if the problem is unbounded. One might argue the man is inherently more able to think and act against uncertainties because the human mind is built to study real-world systems, while the computer is built to study bounded abstract models of systems created by man. I am, thus, a believer in the systems researcher using computers as tools and am quite concerned by systems research activities bounded from the beginning by the limitations of computer models and capabilities.

For those diving into the realm of "Big Data" with terabytes and even petabytes of information, I wish to add a reminder that data is not a mirror to the world, and all large data sets have errors. Errors might occur as a result of the processes in collecting, storing, and transferring data as well as in generating metadata from source data. These errors are typically systematic, occurring in a predictable manner, and can often be corrected through process changes when identified. Errors might also occur through a variety of external factors independent of process, such as random human mistakes in data collection, unforeseen environmental influences, and unanticipated glitches in the mechanistic activities of data management. These errors are typically nonsystematic, occurring in a perceptively random manner. This implies that their detection and correction must often occur in a one-by-one manner. Sometimes, a lack of validating methods might require data mining techniques to be adapted for error identification. Given the size of databases, the challenge is to figure out how to get machines/computers to learn the causes of errors through iterative discovery.

The nature of errors in data includes incorrect information, false information mixed into valid information, missing information, and inconsistent information. Incorrect information can be caused by the data capture person or device (collectors), states and behaviors of the source, and corruption after data capture. False information can be caused by the collector's inability to discriminate/filter data, opposing forces generating false data, and extraneous data that made their way into the database. Missing information can be caused by flawed collection such as not enough range or repeat cycles, flawed transportation such as packet loss across a communications circuit, and flawed storage such as ineffective data architecture design. Finally, inconsistent information can be two or more competing data elements for one parameter, two parameters with a common data element, and data elements in the wrong places.

In this section, I started discussing dynamics within system parts, but dynamics also contribute to the other system formation characteristics that we are about to explore. Therefore, this book is cumulative in its presentation style—each section becoming a foundation stone to understanding following sections. With the dynamics of the parts, the next logical step is to understand how parts associate with one another to form integrated dynamic properties.

2.2   Associations: Connections Between System Parts

As systems are parts working together, there must be associations between the parts. Every part does not have to be and should not be associated with every other part. However, when two parts have an association, there are characteristics that are identifiable. We can deduce four ways to describe the association, as shown in Fig. 2.12: (1) type or types of association that is between the parts; (2) the orientation of the association that can be connected with the orientation of each part; (3) the quality of the association that can be driven by each part's dynamic characteristics; and (4) the effects upon the association that can support or challenge the quality of the association. Beyond these descriptive characteristics, the parts connected together by associations can be treated as a high-level entity/subsystem with total dynamic characteristics as described earlier. In other words, the parts together have group orientation, motion, inputs and outputs, and composition.

In regards to the types of associations, two parts can be linked together in a way that there is a continuous ability to transfer force, exchange physical material, communicate information, and/or detect one another's role and status in the system. For example, the planets in our solar system are continuously under the gravitational force of the Sun and one another. A pipeline could pass oil or natural gas from a reservoir to a user machine, such as a home furnace. A wireless network

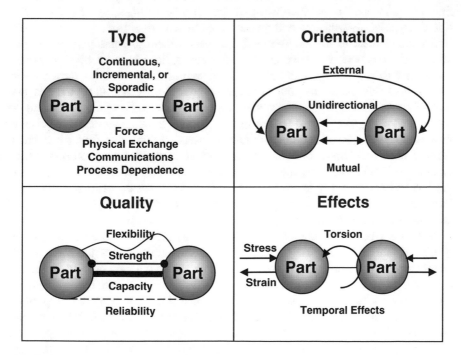

Fig. 2.12 Describing association characteristics

could pass data from one computer to another. And a factory worker could start his/her work, which depends on other people's work, by merely following the directions provided and looking at the clock. In the latter case, the workers are linked together by a process and do not necessarily have to communicate with one another.

The links between parts do not have to enable continuous force, material, and process data exchange. Incremental exchanges are still definable as links. At which point, the nature of the increments, such as time between engagements and duration of interactions, become additional characteristics. Further, the link could form only as needed and break in a sporadic manner depending on the dynamic state of each part. Human players working as a group and adapting to an unpredictable and changing environment will often form and break ad hoc associations as needed. Based on the number of link types and exchange types, there are 12 combined types of associations between parts. Two parts can yield more than one of these 12 association types, and the types can change as a form of dynamics. Sporadic links can suddenly become continuous, and incremental links can gradually become more sporadic. The flow of material between parts could stop while the communications continue, and the communications could stop while the process that associates the parts endures.

If there is any kind of link between parts, the flow of force, material, information, and/or awareness can be in just one direction or both directions. Unidirectional associations often create a subordinate relationship between parts. Beyond these link orientations, a link between two parts can also be external. In other words, the forces, material, information, and process controls can pass from one system part into some element of the environment that passes them to some part in another system. For example, when we are on the Internet, there is not a direct communications link between our computer and the web application server on the other end. Instead, information on both sides are sent out in packets with protocol layers that tell routers throughout a vast global network on how to direct the movement of the packets along adaptive paths. The optimization of the Internet than controls the flow of countless packets so that millions of users all think that they have direct lines of communications. Looking at the Internet example, we can also consider external link orientations as mostly dependent links on system controls, as it is difficult to imagine a simple environment redirecting the paths of forces, material, information, and awareness. In contrast, direct links can be facilitated by a simple medium, such as pipe, wire, paper, etc., or no medium at all.

With link type and orientation, the next set of characteristics is for describing the quality of the link. The quality characteristic of flexibility is how the link can be changed while still remaining intact. The change can be initiated by changes in the dynamics of one or both parts, or it can be due to changes in the medium or mechanism that enables the link. For example, parts that are moving or changing in properties might require the link to also change in order to be viable. If the medium enabling a link, such as air for the passing of sound, water for the shipping of containers, and wire for the passing of electrons, is damaged or altered, then the link characteristics might have to flexibly respond.

The quality characteristic of strength in a link is simply how much opposing force can be withstood by the link based on the reference frame of the link. The opposing force for a link based on forces or physical exchange could act directly on the forces in the link or transport mechanisms enabling the link. Alternatively, the opposing forces could act at each end of the link on the parts. The opposing force for communication links could be false information; the opposing force for process links could be opposing processes.

The quality characteristic of capacity is simply the level of throughput for each type of link. How much force is applied? How much material can pass at one time? How many bytes of data can be transmitted per second? Links based on process might not have a capacity, but the number of procedural steps enabled by the link per second in the process could also be interpreted as capacity in some cases. Finally, the reliability of the link is slightly different than strength in that reliability accounts for the effects of time. A link can be strong initially but become unreliable due to the course of change. Also, the association can simply be unreliable from the beginning as an inherent quality. Not the same as sporadic links, which are known to break and reform, unreliable links fail to meet their objective level of endurance against time and forces. In designed relationships, reliability can sometimes be improved like strength and capacity. Ways to improve reliability include increasing strength, maintenance, backups, and redundancies.

The last set of characteristics for the association of parts is the effects of time, environment, and other parts on the association. Using a physical mechanistic analogy, the effects on an association can be regarded as: (1) stresses; (2) strains; and (3) torsion.

Stresses are things external to the association that seek to force change in the association. In mechanics, stress is the force that presses inward. In a communications link, external stress can be other external communications causing the intended communications to be more difficult. In a physical exchange link, external stress can be checkpoints and detours along a road that chokes the shipping of goods. In a process link, external stress can be other laws and regulations that the process might have to comply with. In mechanics, strain is the force that seeks to pull apart the link. In a communications link, external strain can be parts moving further apart and into areas where the communications infrastructure is less robust. In a link for physical exchange, external strain can be a breakdown of transport vehicles and reduction of delivery personnel. In a process link, external strain can be people unwilling or unable to conform to the process. Finally, torsion is a force that seeks to change orientation. In the case of links between two parts, external torsion might flip the direction of forces, communications, and physical exchange. In a process, torsion might shift the flow of the process. The reason I have added torsion is as a reminder that external forces can affect association in complex ways.

The study of associations given the general characteristics described is separable into the understanding of existing associations and the prediction of future associations as well as changes in associations. In the study of current associations, we first start with whether there are deterministic ways to describe the association— ways that explain cause and effect and ways for the capture of absolute behavior. At times, deterministic definitions of associations are possible because associations are not as diverse and complex as parts. To be specific, the ways to describe parts are unbounded. System parts can be almost anything and anyone, and human components in a system are beyond complex, almost impossible to adequately define. In contrast, the association between even the most complex parts cannot be as complex as the parts themselves. Instead, a subset of the parameters that define the parts will govern the association. Thus, if we can isolate the subset parameters and isolate the nature of the linkage, we might be able to formulate a behavior for the association.

An example for deterministically defining an association is when the force, physical exchange, and/or communication that join two parts follow a wave pattern. Wave patterns are describable by formulas, and the cycles, peaks, and contours can match the association characteristics. Some common wave patterns for incremental links include square waves, sawtooth waves, rectified half waves, and surge and release waves, as shown in Fig. 2.13. Square waves reflect a cyclical flip between two states, such as times when products are moving from Part A to Part B and times when there are breaks. This cycle is definable as a discontinuous function. Sawtooth waves reflect a cyclical rise and decline with discontinuous changes at peak and nadir. This wave/cycle can be used to describe interactions between Part A and Part B that reverse directions based on an upper and lower limit. For example, Part A pours water into Part B until a certain level, and then Part B pours the water back. The link overtime will look like a saw tooth. Rectified half waves reflect a cyclical pulse and a corresponding rest portion. This wave cycle can be used to represent associations based on periodic events with ramp-up start and ramp-down end. Finally, surge and release waves reflect a buildup to a point of sudden release like a capacitor. This release can be periodical discharges of energy that go from Part A to Part B to create an association.

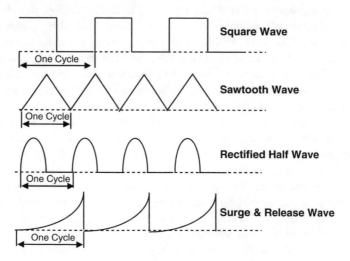

Fig. 2.13 Wave form types for incremental links

Continuous links between two parts with changing characteristics are better expressed as continuous wave functions. These waves can vary in amplitude and cycle overtime, and the patterns of variations for links transmitting energy have been the basis for modern coded communications. In communication links, where a large amount of data encoding is desired, we want shorter waveforms and higher frequencies for modulation. In the links of forces, such as planets in elliptical orbits around the sun, the strength of the forces can vary by distance and other factors to create very long waves. Continuous links do not always have to vary overtime. Instead, the characteristics of the parts joined by a link could change. Thus, there are link dynamics, part dynamics that drive link dynamics, part dynamics, and part dynamics that are shaped by the links.

The types of associations that lend themselves to precise determination of behaviors are, in many cases, a part of designed systems. In natural systems, the associations might be difficult to identify and difficult to quantify. For a particular type of association, the qualities might vary dramatically across a large number of samples without clear causes. Given a defined sample set, we might nevertheless be able to determine the probability of each level of quality occurring and graph the probability of all the quality levels, as shown in Fig. 2.14. If the graph shows a standard probabilistic distribute curve with a dominant mean and most samples falling within the standard deviation, then we are led to consider that there is coherence in the effects upon the association type. If there are multiple clusters of peaks, then we are led to consider that the association type is split by other characteristics and effects to include shifting behaviors over time. For example, people taking a test to become employed in an organization/system will have varying scores depending on individual capabilities. However, we should expect the common test and people's common understanding of the questions to create the coherence and standard distribution of results. The test can then be viewed as a process link between people and the organizational system. If the test is poorly written so that half the people will read the questions differently than the other half,

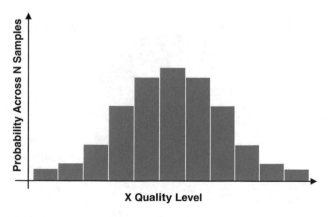

Fig. 2.14 Distribution of probabilities for a coherent association

then the results will be split and the process will be broken. This initial under-standing might help us find linkage issues prior to full understanding and look for the interrelationships between the characteristics of associations to discover causes and effects.

In associations between parts where there is a known group of mutually exclusive types, a pie chart, as shown in Fig. 2.15, can be used to show what percentage of the total number of associations is dominated by each type. The percentages can then be determined by studying large data sets. While we may not have enough understanding to determine which type of link will result when a new association forms, the distribution will provide some insights regarding occurrences in large samples. For example, an organizational component may have multiple choices for sending packages to another organizational component, such as by government mail, commercial shipping, private courier, or employee tasking. While the decision process for selecting the link between sender and recipient at each occurrence of sending a package is highly complex, if the components and con-ditions remain relatively stable over a time frame in which a sample set is estab-lished, then we can build the pie chart. If a shift in the components, such as arrival of new decision-makers, or a shift in the association type, such as price changes, can be specifically identified to enable a new sample set to be collected, then we can study dynamic changes in the association.

If different associations in a system have common but not well-understood characteristics, then the overlap of characteristics between different associations are expressible in a Venn diagram, as shown in Fig. 2.16. This visualization of simple set dynamics can help us focus on the most import groupings of characteristics in a complex system. While the overlaps between three associations are quite simple to see, diagrams involving many associations and sets of characteristics might be visually complex. In this complexity, the human mind might be able to see centers of gravity, hidden stresses, and driving characteristics. If the characteristics being explored also have probabilities of occurrence, then the diagrams can be connected to probability theories for examining the system.

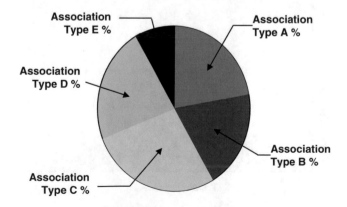

Fig. 2.15 Pie chart breakdown of association types between specific parts

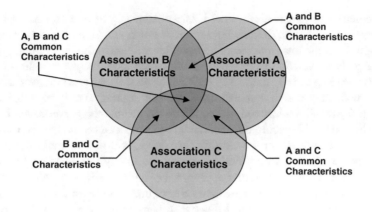

Fig. 2.16 Venn diagram of association characteristics

The use of probabilities to define associations is an acceptance that individual behaviors of elements in a sample population are too difficult to measure. We know that there are commonalities in the elements that drive the determination of probabilities for given behaviors/associations, but we do not know exactly how an element will choose to behave at any given point. The most obvious elements of this type are human beings with our incredibly complex mental processes making decisions based on past experience and thousands of environmental factors. Despite this complexity, people as a whole are remarkably predictable when the population state is measurable and the environmental stimulus is quantifiable. The modern advertising industry is completely based upon this fact. When a test market group of people that is traceable to the profile of the general population is shown a commercial, their probability of reacting in specific ways and forming specific associations are translatable to the overall consumer population. The success of population studies by statistics has extended to medicine, city planning, organizational management, military strategies, and all other missions of human society. The success of statistics has further been extended to other types of populations such as microbes, wild animals, and even environmental phenomena. However, if any of these populations have internal structures of associations/links, statistics will offer limited understanding.

Giving up on understanding how parts in a system will individually behave in achieving associations is contrary to the idea of studying how systems form. Statistical descriptions of association behaviors do not yield true system models if one does not believe that there is true randomness in system behaviors. Even for systems with massive amounts of associations that seem impossible to individually measure, the exact outcome of every association/interaction are predictable if all the conditions in the system and surrounding environment can be modeled. Thus, the notion of a stochastic system can be viewed as merely an abstract concept for coping with the inability to model behaviors or the impractically of modeling individual behaviors. Regarding impracticality, what if the actions of that one specific person or part matter? What if one relationship or one broken relationship can change the outcome of the whole system? For example, a drug company might only care about the total number of acceptable fatalities in the user population for a new drug. However, if the few people with fatal side effects include promising scientists, leading artists, and/or key political leaders, then the whole world is in the balance with only a few associations. I am not arguing about the value of new drugs, and we all understand risks. However, from this perspective, is it not worthwhile to see if specific associations/interactions can be modeled through new explorative techniques? In the case of new drugs, is it not valuable to be able to isolate people who would have adverse reactions through systems analysis?

When deterministic definitions cannot be discovered, and when stochastic assumptions are not enough, there is a series of uncertain reasoning techniques that can be used to explore associations between parts in a system as well as between parts and the environment. Some of these techniques are discussed below, and most of these techniques are being actively explored in research communities. Again, I cannot do justice to these techniques for which researchers have devoted years of effort in developing algorithms and computer models. However, using the concepts for advanced techniques might at times help even novice problem solvers think about the world with its many parts, associations, and systems.

2.2.1 Discovery by Fuzzy Logic

This technique, as shown in Fig. 2.17, treats variables as vague but not random. A fuzzy variable can belong to different sets with varying levels of enclosure. The

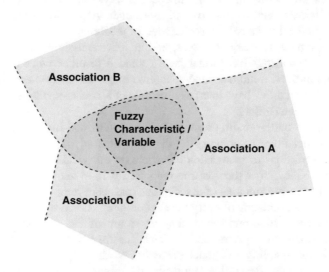

Fig. 2.17 Fuzzy logic connection between characteristics and associations

boundaries of these sets are, therefore, elastic, and the logic for determining enclosure is more driven by degrees instead of simply true or false outcomes. Algorithms with binary outcomes (i.e., If X Then Y) are, therefore, not well suited for fuzzy logic analysis. Instead, fuzzy logic algorithms allow outcomes to have degrees of truth (If Very, If Somewhat, If Not Much, etc.), and the degrees can shift with iteration against changing influences.

The concept of fuzziness has been mathematically explored throughout the twentieth century, but the terminology of fuzzy logic did not become formalized until 1965 [10]. With advancements in computer technology, fuzzy logic has been used to try to mimic complex processes, such as in artificial intelligence, and cope with complex interactions, such as controlling unstable objects. However, at the fundamental level, fuzzy logic is about figuring out associations that are not fully defined but not random in nature. This is most definitely true for interactions based on the human mind as well as all the countermoves to match shifting environmental forces on unstable objects. In both cases, mimicking or modeling outcomes might be easier than figuring out specific associations. Therefore, we might need to find a simpler example of how fuzzy logic can be used to understand associations as a step in understanding how systems form.

That example could be undetermined protein activities as parts in a cell. We know that many complex proteins have a multitude of functions and associations in the course of supporting cell dynamics. When these associations have been isolated, the protein's role in the system can be simply mapped. However, what should we do with protein characteristics that are not clearly tied to associations? Iterative analysis using fuzzy logic and many instances of data might gradually enable us to narrow down the nature of the primary associations (characteristic sets) connected with the uncertain protein characteristic. If we put aside the purest view that fuzzy variables must remain fuzzy, then this technique could move understanding from degrees of belief to near certainty.

2.2.2 Discovery by Bayesian Networks

This technique, as shown in Fig. 2.18, uses a node and link network to propagate a behavior, such as the formation of associations, based on the probability of each uncertain step in the propagation moving forward in one direction or another [11]. The use of probability in this case does not imply automatic acceptance that the entire system is stochastic. Instead, the probability can be used to describe the uncertainty as patterns of belief. These patterns can provide us with insights regarding the accumulated consequences of formed associations. After multiple steps in propagations, we might see propagation paths cross one another, initially dominant paths shift into obscurity, and unrealized ranges in outcomes. The theory in studying associations with such networks is that simple courses of action might lead to complex results over many steps of interaction and association formation.

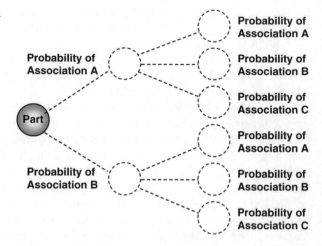

Fig. 2.18 Bayesian network projections of association propagation

An example of parts in a system dealing with a limited course of actions is people in an organization starting to work together to confront a specific crisis situation. The organizational environment and operational procedures provide some constraints on the ranges of people's actions. The emerging crisis creates a degree of uncertainty for people's actions. And different types of people will have different probabilities for choosing the ways to work with others as the crisis progresses. The result is a natural network diagram of associations and evolving associations. Associations that are temporary or enduring can be uniquely identified to portray possible end states. Similar network diagrams can be used to track the status of one person in moving from one action to the next. In either case, we are not examining large populations and assumed randomness. What we are looking for are configurations, patterns, ranges, and overlaps.

2.2.3   Discovery by Rough Set Theory

This technique, as shown in Fig. 2.19, starts with data, such as association characteristics, which clearly falls within specific sets and data, such as uncertain characteristics, which falls within the region between sets [12]. Then, a reductionist approach is used to determine how to be lenient with inconsistent data and to create rough boundaries. For the approach, information is decomposed down to indiscernible elementary granules of knowledge that can form elementary sets (concepts). Elementary sets can then integrate into compound sets, which are then classified as either crispy sets or parts of rough sets. This operationally intense technique meant that robust application did not start until the 1990s as computer processing capabilities increased. In application, the technique crosses from data mining to machining learning that involves large data sets.

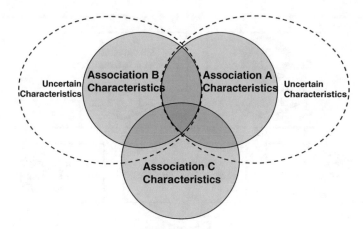

Fig. 2.19 Rough set theory of association characteristics

In discovering associations that are a part of existing and potential systems, rough set theory can be used to filter out system associations in a sea of other types of interactions. For example, in human society, there are countless associations/ relationships between people each with characteristics. Suppose one wants to find out what structure of relationships in the mass of relationships indicates the activities of a terrorist organization, what is the path of discovery? Using this technique, one would take all the characteristics that are from undetermined associations and break them apart, such as a person buying a gun and a person going to the gun store on his birthday. Then, these granules will start to coalesce around association types, such as the person is planning an attack or the person is buying a birthday gift. Granules from other undetermined associations, such as another person interacting with this person and another person buying a gun, might start to shift the set boundaries in favor of specific types associations. When conducted on a set of "Big Data" for population behavior, this type of assessment might produce complex networks of possible associations for indicating hidden systems. In some cases, we may not even wish to start searching for a specific system but, instead, allow the rough set associations to lead us down paths of discovery. Serendipitous discovery can sometimes be highly important in complex system dynamics.

2.2.4 Discovery by Genetic Algorithms

This technique, as shown in Fig. 2.20, leverages evolution theory and uses algorithms that compete candidate search approaches against large data sets [13]. Through comparing fitness parameters connected with the search approaches, a selection using search results is made regarding the best approach to undergo alteration for the next generation of searches. In altering the fittest approach, genetic

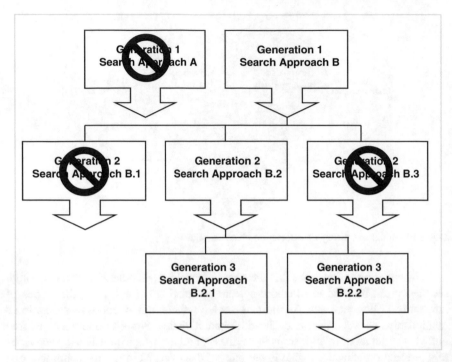

Fig. 2.20 Generations of improving search approaches with genetic algorithms

operators are used to adjust/deviate key instructions in the searches. These operators can also pull positive elements from other search approaches to create hybrid search approaches. The result is a new generation of search approaches based on controlled variation from an evolutionarily successful parent. The goal is to obtain the best search outcomes, even though the initial starting point of discovery may be way off target. In theory, this discovery process will become increasingly intelligent or refined overtime. After generations of searches through complex data environments, one might be able to narrow down the critical pieces of data, which is like finding a needle in the haystack.

Genetic algorithms can clearly be used to search for highly obscure associations without many identifiable characteristics. For example, the early stage propagation of a rare disease with minimal initial symptoms but hugely damaging long-term effects might merit this type of search. How the disease connects with people as hosts, how it proliferates in the host, and how it passes to other people can all be very undetermined at the start of the search. Then, as each victim is discovered, the search can be refined for other victims and more specific propagation paths. This type of discovery is similar to the human process of a detective deductively considering his or her options and pressing forward with each step of an investigation. However, when automated through high-speed computing and sophisticated search evaluation and refinement approaches, the search can quickly advance through hundreds of generations of refinements and a complex path through the data space.

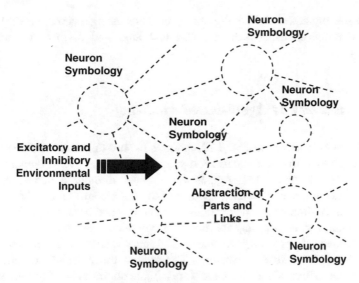

Fig. 2.21 Neural network representation of uncertain association dynamics

2.2.5 Discovery by Neural Networks

This technique, as shown in Fig. 2.21, uses a layered interconnected network of subsymbolic (no rigidly defined symbolic meaning) elements known as neurons to abstractly model a problem space or network of associations [14]. The neurons are defined to have weights as well as connection strengths and can receive positive (excitatory) and negative (inhibitory) inputs from the environment, which induce changes. Therefore, we are attempting to study the dynamics of associations without clear understanding of the parts and the links between parts. After studying the dynamics over a period of time and across a range of conditions, we might then be able to add more specificity to the definition of the neurons and the morphing network structure. Neurons and neuron types that do not fit with clearer definitions can be eliminated. Links that are turning out to be not a part of the system dynamics can also be dropped from the topology. In the opposite direction, increasing dynamic knowledge may cause us to add neurons and links as well as to leave the total topology of the network unbounded. The environmental inputs upon the network that is a stimulus to change could also be initially not well defined. Additionally, the network might have a pattern of change independent of the environment. Deeper understanding of both these change patterns will further contribute to environmental understanding.

Neural networks have been used to study organically formed associations and systems across the World Wide Web. We can track network traffic, identify the ends of communications, and model the dynamics of information flow. However, understanding the people and groups who are communicating and the nature of the communications that form associations can be very challenging. Social networks are often self-organized and rapidly changing. Thus, using a neural network model could reveal the characteristics of nodes and links and discover the extent of associations.

2.2.6   Discovery by Agent-Based Modeling

This technique, as shown in Fig. 2.22, requires the use of an explorative modeling tool that enables us to create agents for representing actors, system components, and environmental elements [15]. Agents are defined through a series of logical behaviors that describe what each agent will do when confronting other agents in the modeling environment. The resulting associations and broken associations over time with many interacting agents might be complex, even though each agent is simply defined. In practice, we want to design agents to be abstractions of real-world entities with the behaviors focused on the associations and system dynamics that we care about. We cannot and should not try to build perfect software replicas of real-world entities. After designing the agents, complex system theory then argues that simple behaviors can interact in complex ways when executed across a large population and/or extended time. The results can reveal hidden patterns, unforeseen consequences, and latent forces.

Agent-based models can be effective at studying organic systems with self-adapting, self-organizing, and self-replicating associations. Obvious organic systems of many actors include ant colonies, bacterial infestations, herd migrations, and other animal groups. However, humans under tactical situations and machine/software codes allowed to freely operate in an environment can also exhibit simple organic interactions. For example, if we use an agent-based model to see how our troops will work with one another and engage enemy troops on a battlefield of

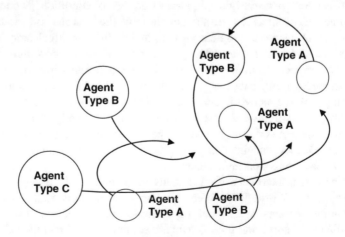

Fig. 2.22 Agent-based model of organic associations

evolving environmental conditions, we might be able to discover new vulnerabilities, unrealized ranges of potential outcomes, and tactical opportunities. The difficulty in such models is that they can only offer ranges of possible futures and not specific projections of outcomes. So the results are difficult to validate against measured data and often not trusted by strategic planners. Yet, when analysts and researchers are confronted with unbound systems and overwhelming complexity, this technique can narrow down the problem scope, help us determine where to find the key associations, and enable us to visualize the dynamics of systems and interacting systems.

In our discussions, I have often used humans and human organizations as examples of complex associations and resulting systems. Therefore, it is worthwhile to discuss the two dominant approaches for simulating human cognition/rational thought. For the past many years, researchers of the human mind have focused on either classical decision-making or naturalistic decision-making.

In classical cognition theory, our rational thought is modeled as a hierarchical tree structure of possible actions and further actions. The theory then argues that the human mind weighs the pros and cons of each course, considers the consequences/outcomes, and determines follow-on actions until the mind discovers an acceptable path to take. In complex situations, strategic planners have tried to organize our choices first by major strategies and then by the tasks for implementing the strategies in different ways [16]. Leaders wishing to follow a disciplined approach for making decisions have enlisted methodologies, such as the Analytical Hierarchical Process, to help them think through the strategies and tasks in a consistent ways [17, 18]. In other words, if they prefer Strategy A over Strategy B and they prefer Strategy B over Strategy C; then the methodology will tell them that they should prefer Strategy A over Strategy C. Classical decision-making can be simulated by computer algorithms as a rules engine that determines what rules to execute based on situational inputs. As such, it could be quite useful in bounded problem spaces and where the choices as well as the logic connected with the choices are clear.

In naturalistic cognition theory, our rational thought is modeled as a comparison of our current situation with similar situations of our past. Using a complete image of a past reference frame or several past reference frames, we then decide what to do in the current situation based on commonalities and differences [19]. The theory is that this process enables our minds to make instant decisions under complex situations, such as soldiers on the battlefield. In support of this model, it can be shown that we can make rapid decisions that are neither random choices nor a path-by-path weighing of options. To simulate naturalistic decision-making, we need to establish a large collection of mental reference frames, rules for how situations are connected with reference frames, and approaches for how decisions can be made based on comparisons.

Regardless of which cognition modeling approach we prefer, it is quiet easy to see that humans are perhaps the most complex and uncertain parts of systems, and systems based primarily on humans without the constraints of clear enforceable rules might be difficult to shape and control. Even the single association of one person to another person can have a multitude of dimensions. So the structure formed by such associations reveals the complexity and dynamic intensity of systems. System structure is, therefore, the natural next step in our exploration of how systems form.

2.3   Structure: All Parts and Associations in the System

Up to this point, I have not really talked about systems formation because I wanted to start with the building blocks of systems: parts and associations. We can see that the dynamics of the parts will translate into the dynamics of the system, and how the system will behave depends on how the parts work together as defined by the associations. Systems can be divided into those formed by man and those formed in nature. Manmade systems, to include systems composed of people, can be precisely designed, controlled in self-formation by people, or uncontrolled in self-formation by people. Because all systems are parts working together, they will each have some type of structure that is defined by specific parts and specific associations. Some systems have common associations from part to part, and some systems have associations that link the parts in step-by-step processes.

In studying system structures, we can start from the individual parts and move up to the total system. Alternatively, we can start with a total system, which has an abstraction of the parts and associations, and progress down to the specificity of parts and associations. If the system is extremely large in the number parts and associations or complex in the behavior of the parts and associations, the abstractions can focus on macro-dynamic behaviors first.

In thinking about macro-dynamic system behaviors, we can envision four types of general system structures as shown in Fig. 2.23: firm and fixed structures, clustered and morphing structures, dynamically linking structures, and dynamically influencing structures. The best way to understand this breakdown is to explore examples, as we will do below. But first, let me say that real-world system structures are not always cleanly divided. Some systems are actually a system of systems with different types of structures at each level and/or region of the overarching system. Some structures may be hybrids, exhibiting qualities of multiple types. These realities raise the questions of system boundaries and system content; both of which we may not fully understand in initially studying a system's structure. The next section is devoted to discussing system boundaries. In regards to system content that are not deliberately simplified in models to study macro-dynamics, there are two strategies for finding missing parts and associations.

The first strategy for content analysis is to assume that the missing parts and associations reside within a bounded section of the system where some level for macro-dynamic behaviors can be measured or modeled. This bounded section can then be treated as a control volume, as shown in Fig. 2.24, with defined inputs and outputs [20]. Based upon matching the inputs and outputs as well as taking into account any known parts in the control volume, research efforts can then theorize

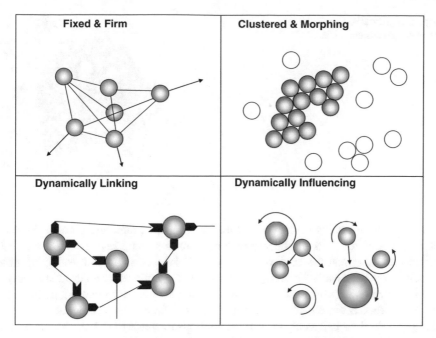

Fig. 2.23 Types of system structures

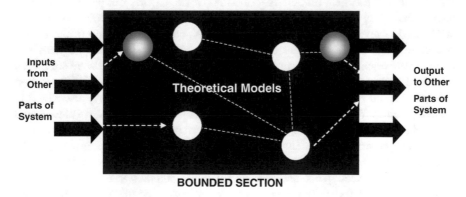

Fig. 2.24 Control volume analysis of unknown system content

and test options for structures inside the control volume that will meet the input and output profiles. In this strategy of bounding the unknown, highly complex input and output profiles make it challenging to formulate theoretical models. However, successful models will be better validated to garner higher confidence.

The second strategy for content analysis is to look across the known parts and associations of a system to see if there is a void, as shown in Fig. 2.25, where links should go into. Indicators of this void could be missing functions and even missing

Fig. 2.25 Black box analysis of unknown system content

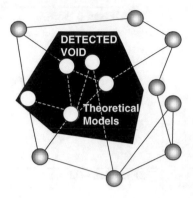

parts. Once we believe that a void might exist and that there are possible links or starting points for links on the surrounding parts for going into the void, research efforts can then treat the void as a "black box" for theorizing and testing possible structural options inside the void [21]. In this strategy of finding holes in the known, the larger voids in densely populated systems are the easiest to find but perhaps the hardest to theorize in regards to internal structures. Small voids could, alternatively, wedge themselves between parts and associations to create perceived disconnects in the associations. Instead of perceived broken links, maybe a void has slipped in.

We need to always keep in mind that the system we perceive may be quite different than the system that exists in the real world. Clearly our perception might have missing awareness of content that hinders our understanding of system structure. However, even with defined structures, our understanding of the details within the parts and associations could reach limits. Doctors a thousand years ago studying the organs of the human body did not understand that they were composed of cells. Doctors a hundred years ago studying cells, discovered in 1665 by Robert Hooke, [22] did not understand how they were controlled by DNA, discovered in 1953 by James Watson and Francis Crick. Even when we think we understand everything within a system, we may not understand all the forces, energy, substances, and communications interacting with the system. This is because not all the associations/links made by parts are necessarily with other parts in the system. Associations between parts and the environment and between parts and other systems could greatly affect system structure and system dynamics. Such associations are specifically discussed in the following sections on interaction and integration.

Instead of getting too far ahead, let we define the four types of system structures and let us explore examples that explain the merit of each definition. People have devoted entire lifetimes to studying each of the example systems, and each course of study has resulted in specific theories, models, and formulas. Our exploration at this stage is not to debate the current research or use current research to control specific systems. Instead, let us think about how systems are similar at conceptual and structural levels before we start to study how systems are unique.

2.3.1 Fixed and Firm Structures

System structures can be considered fixed if the associations between parts endure for extended periods and the characteristics of associations remain unchanged. These structures are then firm if they can maintain their fixed configuration under internal and external forces. The limits of this definition depend on how long the structure has to be unchanging in order to be considered fixed and how much force the structure has to resist in order to be firm. Typically, in order for the structure to be fixed, the associations have to be well-defined links between set parts. One way for this definition to emerge is if the links between parts form a connected process. This process then creates dependencies between parts and the strength of each part's ability to sustain its link impacts the total firmness of the structure.

Structures can have all common associations, which often means that the parts are also similar. In such cases, the dynamics of the system resides primarily in the changes moving across the links and not with the characteristics of the links. For example, all the links in an electrical power grid delivering electrical current from power station to power station are similar and the system looks static from the perspective of the infrastructure. However, the system is dynamic because of the power that flows through it. In contrast, the parts of a mechanical clock are all moving in perfect synchronicity. The dependencies and the dynamics are all visible and clear.

Many fixed systems can have highly volatile dynamics within the links and the parts/nodes. Parts in fixed systems typically cannot change orientation or motion independent of other parts. For example, all the gears of a clock must turn together. And parts typically cannot change the type of inputs and outputs as well as their internal structure and surface features. However, parts can accept changing levels of force, energy, substances, and communications. If these changes in the connection of the parts are too fast or exceed maximum or minimum limits, the internal stress can cause a part to break away from the system, reorient, move, or compositionally change in a way that causes fixed links to break or change. The continuous links in a strong fixed system should resist stresses and strains, as discussed earlier. The ways to resist are to have strong connections in how links join parts; have enough capacity to sustain the dynamic flow of forces, energy, substances, and communications; and have high reliability. Nevertheless, internal dynamics can place great stresses and strains on the structure just as external forces can push and pull upon the nodes. Systems that look very firm can, therefore, be on the brink of sudden collapse.

Fixed systems can expand in size by adding parts and links to the systems. This expansion, however, has to be a planned build or growth event in moving from one fixed state to another fixed state. If the expansion is continuous, then the system is not fixed. In a build event, parts are either acquired from outside the system or constructed by the system using external supplies. If the system has to use salvaged resources to build parts, such as other failed parts, then the structure is not very fixed. In a growth event, parts can self-generate new parts either by division or reproduction. Either way, external inputs will be required to sustain the growth. Noncontinuous growth usually occurs according to fixed cycles. At each cycle, the parts can grow throughout the structure, in specific regions or around the surface. Following the cycle, the structure then goes through a transition period where new links are established and system structure is extended.

The following examples reveal the prevalence of fixed systems in our natural and man-made world. In fact, these systems are so common that we often overlook their constrained system dynamics until something goes wrong.

The Body as an Example of Fixed and Firm Structures: We, human beings, are organic systems with fixed subsystems centered on primary organs held together by the skin and skeleton of the body. The subsystems include respiration and circulation enabled by the lungs and the heart; digestion enabled by the stomach, liver, and intestines; thought and sensation enabled by the brain and sensory parts; movement enabled by muscles and tendons; liquid waste management enabled by the bladder and kidneys; and reproduction enabled by the sex organs [23]. Each part of the subsystems is composed of building blocks called cells, which are by themselves complex fixed systems. From the perspective of the body, the subsystems and their organs work together with complete interdependency, and the system will break if any organs/parts fail. The organs interact with one another through the passing of red blood cells (erythrocytes) to carry oxygen and remove carbon dioxide, nutritional substances directly through cell membranes, protein structures for intercellular and intracellular communications and control, and hormonal type chemicals like adrenaline for organ level control. Further, the nervous subsystem uses chemically driven electrical impulse for thought, organ regulation, and control of the overall body system. Although this is an extremely simplified description of the highly complex human system, it clearly shows how a fixed system can succeed in nature. The complexity of nature allows this type of fixed system to vary greatly across different types of animal life and even more greatly between animal and plant life. The complexity of nature further allows the human system to vary slightly between person to person to allow for diversity and individual success across the human species.

The Cell as an Example of Fixed and Firm Structures: Cells are the building blocks from animal and plant life and represent the total system for single-celled prokaryotic life in the categories of bacteria and archaea and eukaryotic life in the category of protists. Cell structures vary greatly between prokaryotic cells, which do not have a nucleus to protect the circular DNA control structures, and the eukaryotic cells of animals and plants. Animal cells, with soft membranes, further varying greatly from the firm structured plant cells in their internal chemical reactions to sustain life. Finally, animal cells even in the body of one animal will vary greatly to serve different functions in the body.

Nevertheless, each cell is a very fixed system, which only changes according to the mitosis reproduction cycle across the S and G2 phases of cell life. However, this cycle is slowed or suspended for cells in bodies that have achieved adult status where further growth in size is no longer necessary. Bacteria and archaea cells are less stable in that they are continuously reproducing through mitosis and expressing mutation effects from environmental chemicals and radiation as well as from the DNA of other systems. The parts of an animal cell include the DNA blueprint organized into linear chromosome molecules, nuclear membrane that protects the chromosomes, ribosomes with RNA structures that support protein synthesis, mitochondria that support energy generation, vesicle for material transport and storage, and Golgi apparatus for protein packaging. As noted, this discussion is not to make us cell experts but to show how nature has been successful in designing fixed systems. In the case of the cell, the system can enable the formation of larger more organized fixed systems like the organs of the body or it can enable single-cell life that work together under other system constructs to be discussed.

The Government as Example of Fixed and Firm Structures: It might seem quite strange to jump from human physical systems to government systems, but the nature of people living together is to start forming structures of leadership and control. The structure of government from the earliest tribal chief is perhaps the oldest human-created system. Since the goal of government is to sustain a stable society, government structures should be fixed and firm. The simplest way to achieve a firm government is to crown a king with central control authority and designate appropriate nobles to manage the affairs of regions. Because absolute power can be abused, the monarchical structure of government can be constrained by constitutional laws, such as the Magna Carta [24]. In complex societies, in which the people are educated and have achieved economic prosperity, the people may demand participation in government through democratically elected representatives and direct approval of key decisions through a popular vote. The goal of democracy from our perspective should be to sustain a firm system of government where leadership change does not collapse the system.

Regardless of the specific form in government structures, all stable governments are a fixed system with typically hierarchical structures of responsibilities. The management can be based on regions, core societal services, population groups, or a combination of all three. However, the links need to be clear for those in government and those being governed. The society being governed can, in contrast, be a dynamic entity with economic fluctuations, introduction of new technologies, and migrations of the people. The society can alternatively be simple with one core marketable product and all other resources acquired through trade. Regardless of the dynamic properties of society, the firm government system must integrate into it to exert effective control. The challenges of system integration will be discussed in a later section. However, I must note that effective control has been a topic of political and economic theories and debates for many years.

Military Forces as an Example of Fixed and Firm Structures: Military force structures emerged almost as early as the first government structures. Force structure refers to the organization of people to fight as a coherent group to increase total combat power. This includes recruiting, training, a hierarchy of command, equipping of soldiers, procedures for roles in the structure, a process for promotions, and capability to treat the injured. In modern combat, the soldiers must adapt to the chaos of the battlefield, adversary strategies and tactics, environmental stresses, and the risk of unforeseen events. While the coordinated behaviors of soldiers could resemble other types of system structures to be discussed below, the dynamics of war actually emphasize the importance of maintaining a firm force structure, a clear way to deal with chaos. When parts are damaged and associations are broken, there must be a way for the system to repair the structure. If new parts and associations cannot be established, the system must have a way to collapse down to another firm configuration. The reference frame in which a military force structure is fixed is not the physical terrain but the information terrain. Soldiers in combat could suddenly be hundreds of miles away. But as long as the communication links are maintained and all sides recognize the responsibilities, the associations and parts are firm. The military can consume supplies in a variety of patterns depending on the status of conflict. But as long as the supply lines are maintained, then only the flow is changing across the links. Based on historical examples, a well-organized military force structure might be able to maintain its stability even after losing more than 50 % of its parts (soldiers and weapons) [25]. I will return to discussing the military in exploring system interactions and how systems fail.

Companies as an Example of Fixed and Firm Structures: For-profit companies are societal structures organized by people or the state to gather wealth through integrated work activities. These work activities can lead to products, services, and/or the manipulation of existing wealth. The integration of work activities is typically based on established processes and procedures enabled through a firm system structure. In highly volatile markets, processes can be designed with flexibility and responsiveness. However, the structure of the company should still be firm to anchor the processes. For example, select people in a company may be given great latitude to make individual decisions against market forces, but these people must still recognize their reporting chain and who can disapprove of their efforts. Otherwise, there will be no alignment of processes, control of integrated work, or reliable associations to sustain the system. A bunch of people working successfully each in their own way does not automatically make a company. A company must have a structure of control. This control can be purely hierarchical with the hierarchy to include positions such as president and CEO, C-Level officers, division vice presidents, directors, managers, and employees. Alternatively, this control can be a matrix of horizontals, such as with regional vice presidents in charge of markets, and verticals, such as vice presidents in charge of product lines and capabilities being delivered to the markets. Further, the company control structure can be very flat, with few layers of management and a lean organizational structure. Alternatively, the company can have extensive controls across layers of management for enabling continuous oversight. The best or firmest (most resilient) structure for a company, therefore, depends on what is being sold, how things are produced, and the market forces.

Educational Systems as an Example of Fixed and Firm Structures: Parents have been teaching children and masters have been teaching apprentices since the first tribal societies. Formal educational systems, with structures of management, professors, and levels of students, emerged in Europe based on the Cathedral Schools of the Middle Ages [26]. Though the medieval schools were to serve the nobility and to supply churches with educated clergymen, large educational systems would later emerge across the world to supply companies and the government with skilled workers in the industrial and information ages. The educational systems at the grade school and high school levels have become prevalent across modern societies because all workers need a core body of knowledge and skills. Then, the students can proceed to universities, colleges, and trade schools to prepare for specific career paths. The structures of the educational systems are so fixed that many have endured for centuries even as the courses taught are undergoing continuous change. Pictures of generations of chancellors are hanging on university walls. Statues of famous university affiliates are occupying campus lawns. And stories of university accomplishments are passed to the student and alumni population. Walk into any school today, from high schools to universities, and one will see the instruments for strengthening system structure. The system instills traditions, recognizes the best in each generation, enforces links and paths, celebrates group accomplishments, and welcomes back those who have graduated.

Mechanical Machines as an Example of Fixed and Firm Structures: The design of machines is man's commitment to the power of fixed systems. From the creation of the wagon to modern day jet planes, machines designed by man have parts that must be made precisely, assembled into a physical system, and work synchronously to achieve the intended dynamics of the system (i.e., to transport goods along roads, fly people across continents, and manufacture goods). Some machines, such as robots, can have a wide range of motion and functions. What is fixed and firm in the structure of machines is the association of the parts to one another. Even if a part can change associations based on operating conditions, that ability to change must have been designed into the mechanical system. To preserve the fixed state of machines, a process for replacing aging or broken parts is often incorporated as an element of the design to extend system life. In more and more cases, sensors are further embedded into machines to detect part failures. Current mechanical systems cannot self-replicate like organic systems. There are no ways for system components to gather more raw materials from the environment and the mechanical systems still cannot fashion raw materials into parts to build copies of themselves. 3D printing technology might move machines toward self-replication [27]. For now, all machines have a life cycle, and all machines will need to be retired unless so many parts have been replaced that one is getting a new machine. How machines fail will be discussed later in this book.

Infrastructure Systems as an Example of Fixed and Firm Structures: Modern human societies are sustained by electrical power, water supply, sewage, fuels, fiber communication networks, and broadcast networks. Our homes literally sit on top of these infrastructure components, which consist of mechanical machines, computers and electronics, devices such as valves, and nondynamic parts that facilitate transport. As I have mentioned earlier, infrastructure systems, excluding orbiting telecommunication satellites, can appear static in the physical world. However, the dynamics of energy, material, and information flow across the system are what make infrastructure a system. The firmness of the system or resilience against disasters, such as hurricanes, earthquakes, bombs, and blizzards, depends on (1) the system's ability to resist or avoid damage; (2) the system's ability to be repaired once damaged; and (3) the system's ability to reorganize flow to minimize the impact of damage. Modern communication networks with dynamic routing of information are highly able to reorganize information paths. Well-managed city departments have crews that are ready to repair downed power lines and broken transformers. And some water supply and sewage systems are designed to redirect floodwaters. As people and systems used by people in urban environments depend heavily on the infrastructure, the failure of this firm system has serious consequences.

Electronics and Computer Hardware as an Example of Fixed and Firm Structures: Systems that depend on electrical supply include a variety of electronic and computer devices. Today, computer processors have become so versatile that most electronic devices from the television to kitchen appliances all have embedded computer controls. When we look inside an electronic device with all its components and wiring, it is clear that it is a fixed system carefully designed to route, amplify, and utilize electrical power. However, when looking at a computer board with all the complex electrical pathways and embedded microchips, it can be difficult to understand what is fixed. The heart of the computer resides in the capabilities of the microprocessor and random access memory (RAM). These and all integrated circuits have millions of tiny transistors inside interconnected by a complex web of semi-conducting material. The semi-conducting material can be as thin as tens of nanometers and still carry electrical current. The microscopic transistors then control the routing of the electrical impulses to enable computing functions [28]. This elementary presentation is only to point out that the designs of integrated circuits are still fixed though extremely complex. In fact, the complex structure is packed so tightly that the integrated circuit chip typically either survives as a whole or fails as a whole. Over the past few decades, computer chips have mostly been discarded as obsolete technology long before they have reached the point of failure.

Software Applications as an Example of Fixed and Firm Structures: The purpose of computer hardware systems is to sustain software that supports all functions in society and most devices that we use. The question, however, is whether software, as merely an interlinked series of commands that operate on data and external inputs, constitute a system. The commands are in the form of a programming language for people to create software. Then, the developed code is either compiled into the binary-number-based language of the computer processors in advance or interpreted in real time as the software is executed. Either way, software is a linear series of codes that can be printed on paper. These codes can be changed by the programmer to correct bugs/programming error, but they are fixed except when codes are designed to self-modify. When stored and inactive, software does not have the dynamics to represent a system. However, when a piece of software is executed by the computer, it becomes a system that gathers and outputs data, organizes and stores data, computes data to create metadata, and controls physical systems based on computed results. This process, which can be highly interactive with users and the environment, will appear adaptive and constantly changing. However, the information is actually moving about a coded system of controls. Parts of the system may never be utilized, but the software of current computers is a system that seeks stability. The level of stability or firmness depends on how easily the codes can get corrupted due to errors in operations, exploited by software virus, and accessed by computer hackers.

Moving on, I have clearly not covered all the countless fixed and firm system structures in the world. Nevertheless, the presented examples have hopefully shown that firmness is often a dominant state but not an absolute condition. Under the right stresses and strains, fixed systems can break or transform into other structure types. The formation of this dominant system structural state is, therefore, about tightly connected processes, dependent parts, and resistance against damaging effects.

2.3.2   Clustered and Morphing Structures

System structures can be considered clustered if the associations between the parts are close based on the primary reference frame in which the system is being measured. Closeness does not have to mean distance but is often tied to the time of interaction from one part to another. This closeness can be caused by the nature of the parts and/or links. However, it can also be caused by the forces that travel across the links as well as external forces pushing the parts together. Clusters can be small, having a few parts as long as the number is more than two. Clusters can also be huge, having millions of parts. A cluster of parts that do not shift their associations with one another is simply a densely formed fixed system. However, many clusters with close associations have enough forces and/or links keeping the cluster together that individual associations can change without breaking apart the system structure. These changes enable the structure to morph as the parts remain together. By this definition, a clustered structural type is the only one that cannot enable a distributed system structure with parts spread far away. A distributed rigid structure will be difficult to sustain and perhaps difficult to redirect motion, but it can exist as long as the links are strong enough. Dynamically linking and influencing structures can be distributed to the extent the links can reach or the parts can travel. This leaves the idea of clustering and distributing as directly opposite concepts.

The ways that associations can enable the clustered system structure to morph is to have flexibility and/or incremental properties. In a cluster, flexible links enable parts to shift positions relative to one another in the structure without breaking the links. The links can reorient and stretch, but there is enough strength and pull in the links to keep the parts together. In contrast, incremental links will break as parts shift, but new links will form fast as the parts get into new positions. The incremental links in a cluster cannot all break at once. Instead, the structure must always be held together by some links as the other links are reforming. Finally, a morphing structure can have a mix of flexible and incremental links or links that are both flexible and incremental. In the latter case, the links will try to stay connect until parts shifts beyond a certain point. Then, old links will break as new links form.

Based on the above definition, clustered and morphing system structures typically cannot have uniquely defined associations and process components. The uniqueness of the links and the specific dependencies across the structure encourage fixed states, thus making redefinition of relationships among many parts difficult. This does not mean that all the associations in morphing structures must be the same. Complex parts might be able to support a variety of situationally dependent associations, and parts with self-control or even self-reasoning capability can build highly complex networks of varying links in morphing structures. For example, people living in close proximity to one another might form a clustered system based on their relationships. This system is different than people in a company with clearly defined roles and dependencies. However, the social structure can still have many types of local associations and highly complex group dynamics. What keeps the structure together are common characteristics in the associations such as the need for friendships, the search for spouses, and the sharing of social resources. Using social clusters as an example, we see that fixed structures with elaborate processes can exist on top of morphing structures such as specific job responsibilities for people in a social cluster. We also see that fixed structures with elaborate processes can be threatened by self-formed associations between the parts that will break the structure down to a morphing system. What if the workers in a factory organize into a social structure that pushes against the defined workflow and conditions? We then see two systems fighting to exist with the same parts or overlapping parts. I will save this discussion for the section on integration.

Returning to the consequences of morphing structures, the volume of a system in its reference frame is governed by the number of parts and closeness of the associations. However, the structure can morph into many shapes because of internal and/or external forces and because of the functions in the system. When morphing is a reaction to forces, there should be an inherent level of resistance before structural changes start. Thus, the structure will be fixed once the forces stop. In morphing as a part of system functions, the parts and links could shift in the course of working together or due to some level of centralized control. The system might morph to evade, invade, engulf, pass through, or block.

Clustered and morphing systems can expand or contract continuously as well as in phases. Expansion can be through pulling parts into the system, creating parts in the system, or loosening the closeness of links. The ability to rapidly form links in these structures enables high rates of growth. The ability to shift links to other parts further allows the system to cope with high rates of parts breakdown. These changes in parts can work in conjunction with the morphing of the total system. In case of system damage, the morphing and growth can help restore weakened regions and overall parts density.

The following examples reveal the immensity of some clustered system structures and the intensity of forces involved in structural changes. Most of these systems are found in nature, as not many designed systems with man-made parts have adopted clustered self-organizing dynamics. This could change with advancements in nanotechnologies [29], micro-robotics [30], and automated swarming weapon systems [31].

Climate System as an Example of Clustered and Morphing Structures: The weather patterns of planet Earth form a complex system that includes effects from the Sun, Moon, Earth's rotation, and Earth's revolution. The gravity of the Moon causes high and low tides at where the ocean meets the land. The thermo energy from the Sun powers the circulation of the air. The position of Earth on its rotational axis and about its orbit yields the seasons. Water from the oceans, seas, and lakes fuels the clouds and transfers thermo energy to create storms from hurricanes to blizzards, and mountains and rivers capture precipitation to feed the oceans, seas, and lakes.

The Earth's climate system is powered by thermo energy along with changes in temperature states. However, water and air are its primary parts. Therefore, if we observe water and air molecules across the planet, we see a massive clustered and morphing system structure that moves and passes force as well as energy through close interaction of the molecules. The system on a molecular level is actually quite simple, even though air is composed of oxygen, nitrogen, carbon monoxide, argon, and green house gases. Green house gases include carbon dioxide, nitrogen oxide, methane, heavy oxygen in the form of ozone, and chlorofluorocarbon. It is the interaction of these molecules across the planet that creates complexity beyond the ability of current computer models to precisely predict. Thus, we get incorrect weather reports and the climate change debate [32].

The debate regarding whether the Earth's climate system is failing to support a stable human society is centered on the increase of atmospheric green house gas that can be traced to worldwide deforestation and burning of carbon fuels. Green house gases are important for maintaining surface temperature, but an increase in these gases that trap solar radiation could cause temperature increases and climate instability. The debate is that we know that human endeavors such as over farming can cause dust bowls and over burning of fuels can cause city smug. However, there is still a lack of direct traceability between increases in green house gases and temperature change. The Earth had gone through many major temperature-change cycles long before human effects; thus, the level of human effects relative to Earth's own climate dynamics is still uncertain. If the effects of man are small compared with the Earth's own climate shifts, the world might still be heading toward a climate oscillation and ice age due to melting polar ice. Unfortunately, in such a case, closing down factories and using clean fuels will not prevent this reality.

Terrain Systems as an Example of Clustered and Morphing Structures: A much slower system than the Earth's climate is the terrain/surface features of the Earth and other solid planets. Nevertheless, forces are continuously moving across the Earth's solid mantel layer that sits on top of the molten outer core. As a result of these forces that pass across the clustered material structures of the ground, mountains rise, valleys form, rivers start, and bodies of water emerge. Sudden shifts between surface plates and the sudden collapse of the surface structures can cause earthquakes. And lava from the molten outer core as well as water heated by lava can burst their way to the surface in the form of volcanoes and geysers.

Even ignoring sudden tectonic disturbances, the surface of the Earth is highly morphing when measured across millions of years. This surface is composed of localized material structures such as rock formations, sand and soil, and even crystalline elements, which can all be viewed as system parts. When we remove a part from the Earth, it is just a simple object. When we look at the Earth in human time frames, it is just a static platform for architectural endeavors. Therefore, we care about this system primarily for the sudden disturbances that result from thousands of years of gradual change. Once an earthquake, volcano, geyser, or rockslide occurs, the power of this morphing system is realized. The power in some cases can be triggered by human error such as incorrect mining practices, poor construction decisions, and deforestation. Therefore, nations, companies, and city planners all have a vested interest in understanding the dynamics of terrain systems.

Bacteria Growth as an Example of Clustered and Morphing Structures: I have introduced bacterium as a fixed single-cell organic system, but the reproductive growth of bacterium through mitosis cell division can rapidly form clusters of bacteria that constitute greater morphing systems. A cluster can invade nearby cell structures in a host organism or dominate a nutrient-rich environment such as food. In the spreading of the bacteria cluster, the association between bacteria cells might be minimal. However, bacterium can communicate with one another through chemical signals known as quorum sensing [33]. In this communication, bacteria in a cluster can coordinate their movement and growth activities to increase their success at overwhelming host's defenses. One key coordinated action is for the cluster to wait until it is large enough before launching a major assault on the host. Another coordinated action is determining which chemical should be produced by all the bacteria. As an invasion progresses, the bacterial growth and spread can be along key pathways in the host. The bacteria can simply eat away at surrounding tissue in the host, but a pathway would be for the bacteria to travel along the blood stream and attack specific organs in the host. Some bacteria regulate their growth to achieve symbiotic existence with the host body. This symbiotic existence can benefit the host as in the case of bacteria in the human intestine that aids digestion. However, symbiotic existence can also degrade the host by taking away nutritional content, continuously battling the immune system, and causing low-level cell damages.

To defeat an invading clustered and morphing bacterial system, one can kill off individual bacterium faster than their growth rate, kill all the bacteria by exploiting a common weakness, disrupt their chemical signals to weaken coordinated attacks, and/or manipulate their chemical signals to break apart the system. In the open environment or on the surface of bacteria-infected wounds, antibacterial substances, such as alcohol or hydrogen peroxide, will kill the cell structures of bacteria. Inside the body, antibiotics such as penicillin can either kill bacteria or inhibit their growth process. However, bacteria evolve quickly. For every bacterium that has been exposed to antibiotics but is not killed, there is the chance that its mutated characteristics will survive to form strains of antibiotic-resistant bacteria. So the story of bacteria is one of competing systems, which will be further explored later in this book.

Population Migration as an Example of Clustered and Morphing Structures: I have described the infrastructure and organizations of societies as fixed systems, but the people in society do not have to be trapped by these fixed systems. In times of danger or in search of opportunities, people have been known to cluster and migrate. This behavior is very different than people gathering to watch a ball game where there is minimal association between people to yield a system. Instead, when people are moving toward an opportunity, such as the wagon trains going to the American west, they band together to help one another achieve a common purpose. The collaboration can be as simple as forming defensive circles when under attack or as complex as the exchanging of vital resources.

When people are moving away from a threat, such as running from a forest fire, they must at times coordinate with one another to prevent gridlock and chaos. The coordination can be as simple as some people directing the flow of traffic and as complex as continuous exchanges of situational awareness across the cluster. Humans have not always worked well together. So this type of system can have very high transformability but very weak cohesion in the cluster. People might start to move faster as panic increases, and the cluster might break apart as individual fears take precedence over group survival instincts.

Mob Actions as an Example of Clustered and Morphing Structures: The natural condition for forming a strong human cluster is, sadly, when everyone shares a common intense emotion, typically anger, and wants to take action. The emotion creates tight bonds between people, then the cluster/mob will morph once one person starts to take action. This morphing might be an assault on a section of the city when people, without centralized guidance, help one another tear down statues, burn buildings, spread graffiti, and vandalize stores. Because of the lack in centralized control, mob-type systems are difficult to confront. People in mobs can still be adaptive and clever in their attacks, and the aggressiveness of the attacks will not dissipate until the emotional intensity dies out. The associations in mobs can span the range of human communications and communication devices. Sometimes, it is the simplicity of communications and the localized rallying of actions that make a mob effective.

Strategies to deal with a mob system include disbursing the people enough with law enforcement or troops so that cluster dynamics break down, eliminating the causes of emotional intensity through negotiations, terrorizing mob participants into inaction through greater violence or arrests, and attacking all the mob participants. Generally, the sequence of events in dealing with a mob is attempts at negotiation and disbursal followed by individual arrests and broad attacks. The latter actions might cause society-wide backlash if other people are also sympathetic to the cause of the mob.

Close Quarter Troop Engagements as an Example of Clustered and Morphing Structures: Successful human clustered and morphing systems are, at times, created through training. A tightly formed group of combat troops is one such system. Even in modern warfare, combat in jungle and urban environments might require troops to work together in a cluster to engage enemy forces. The cluster might not be large, but high-tempo operations depend on coordinated tactics and common understanding of set procedures. For example, communications when enemies are near might rely upon hand signals, flashing codes, text messages, and special sounds. Once the system is in action, the morphing pattern can be to move in one direction, cover multiple paths and angles, surround the enemy, divide the enemy, sneak up on enemy, hide and evade, regroup, or retreat. For highly trained troops that can act independently and fight in coordination, a clustered system can remain effective even under heavy losses. As the cluster components are trained to maintain the cohesion of the cluster, the system is far more difficult to break apart than mobs.

The few examples of clustered and morphing systems presented reveal great diversity among this type of system structure with the power of transformation. Many system structures can morph. However, the morphing of a cluster of system parts is the most obvious. The cohesion of the cluster adds strength to the transformations and the size of the cluster increases the power of the total system.

2.3.3   Dynamically Linking Structures

Systems can also have loose structures that are formed by temporary associations. Such associations are formed by the changing orientation, motion, and/or composition of parts. The surface conditions of the parts, to include input and output interface standards/openings, then governs how the links are established and when the links will break. For example, if the content of a part exceeds a specific level, the internal stresses might force a link to form with another part to transfer content and lower stress. Alternatively, a part might have a content deficiency that compels it to establish links with sources of available content. The key difference between dynamically linking structures and clustered structures is that the links are not formed by proximity between the parts or forces pushing the parts together. Therefore, the links must have functions that keep parts together, but the links can stretch and shift across substantial distances, as long as the functions of the links are sustained.

Dynamically linking structures can cover a vast space and expand or contract significantly within space. Parts can be expelled from the system when links are no longer required and parts can be added to the system when links are justified. Further unlike clustered structures, complex processes can form across the parts and links with high degrees of uniqueness throughout the process flow. In such cases, the uniqueness of the parts and links will control the dynamic behaviors of the system, the rate of system expansion or contraction, and the realignment of the structure. The capabilities of the system enabled by dynamic links can yield highly self-organizing and self-adapting properties. This organization and adaptation can concur through the autonomy of the parts or through centralized control. However, if the system processes across the dynamically linked structure are complex, then the controls within the parts or across the links must match the complexity.

Dynamically linked structures typically have no baseline size. Therefore, system expansion and contraction is an inherent characteristic and has no special meaning. Parts can proliferate or be built inside the space of the system. However, the parts will not belong to the system structure until a new role or repeated role has been established through dynamic association. This definition creates a challenge in establishing system boundaries, which we will further explore in the next section. For now, the simplest way to describe the challenge is the question of: If parts are dynamically linking to one another, how do we know what links are between parts in the system structure and what links are to parts in the environment? The answer will require an assessment of commonalities and differences between specific parts and links for a defined system. Based on the assessment, a practical system boundary can be drawn that does not include every single element/part that interacts with one another.

The following examples show how dynamically linked structures are observable in nature and incorporated into human activities. In studying the structural behaviors, the uncertain reasoning techniques for identifying linkages, as discussed on the previous section, might yield greater insight. We tend to think of dynamics as links forming fast and/or often. However, it might be the slowly formed links that are the most important and most difficult to study. For systems that have slowly forming links, key parts within the system might exist in plain sight but not be revealed for decades. To illustrate, a sleeper agent/spy planted in an enemy's organization would be a key part of an opposing system. Yet that association and role will not be visible until the agent is activated through a triggering link.

Insect Colonies as an Example of Dynamically Linking Structures: Some insects, such as ants, will work together as a distributed system centralized on a nest or hive housing the queen. For ants, the workers then move out from the nest in search of food and building materials. The male ants with wings also fly out to mate with other queen ants. When foraging in the environment, worker and soldier ants can individually wander far away. However, they will still circle back to communicate with one another to coordinate actions. Hundreds and thousands of ants can quickly gather to bring food back along a transport column once a food source has been identified. The soldier ants will organize for an attack when another insect species, such as termites, is encountered. In the nest, the ants will work together to dig caverns and construct living spaces. Thus, the status of the ant colony system largely depends on communications between ants, the immediate needs of the colony, threats facing the queen, and opportunities in the environment. The ability of each ant to act independently and form associations as required is what makes the colony a dynamically linking system. The system is not held together by the links but by each ant/part's commitment to the system. Even when disconnected from the whole, the ant will perform its responsibilities for the system and seek to return to the system. When necessary, worker ants will even sacrifice themselves for the system. This is why a dynamically linking system can still be very strong.

Protein Interactions as an Example of Dynamically Linking Structures: Proteins are complex amino acid molecules organized into polypeptide chains. These chains are produced by organisms to perform a vast variety of functions [34].

Different folding protein structures support functions such as DNA transcription, catalyzation of cell chemical (enzyme structures), intercellular communications/ signal transduction (structures like insulin), cell defense (antibody structures), material transport (specific protein binding sites), and cell integrity (fiber proteins). Thus, the bodies of living things are filled with proteins working together to sustain life. Proteins do not have intelligence but, instead, a complex range of behaviors based on molecular structure, folding dynamics, and surface binding sites. When these molecules are outside of the body, they are merely nutritional elements for the body to use in energy generation and cell construction. Within the environment of the body, however, proteins are responsible for most of the dynamics within cells and between cells. Proteins' associations with one another include passing components and triggering reactions. Yet, the most important association is perhaps a common anchor to the cellular reference frame. In this reference frame, the cells of organisms control the production and types of proteins released as well as the retirement of proteins. It is through this control and the suitability of proteins for specific functions that system cohesion is achieved.

Social Networks as an Example of Dynamically Linking Structures: I have placed computer infrastructure and the worldwide network of fiber optics and cell phone towers as fixed systems. However, the information on these systems, as well as the activities of people in generating, consuming, and manipulating this information, is far from fixed. Generally, a social network is enabled by a web-browser-based software application that allows people to share information based a set of rules and constraints.

The first set of rules governs role-based information access, which is supported by user identity management [35]:

- Who can sign on the application?
- How is the information shared among people with different access privileges?
- What information can be gathered by all participants on the application?
- What information can be used by the application admin for purposes such as advertising?
- What information is searchable on the open worldwide web?

The second set of rules governs the type of information that can be shared: lengthy text, short text, photos, videos, web addresses links, links to content on application, etc.

The third set of rules governs the quality of content such as filters against adult material, filters against political opinions, filters against harassment-type activities, and filters against copyrighted content.

There may be many other rules that govern associations between people, but it is clear that social networks are dynamically linking structures that can grow and shift rapidly across the population. People are joining and leaving networks all the time. The information that people share changes with life experiences, emotional states, and social interactions. The links people make change base on friendships, personal interests, and professional needs. Social networks are spawning new product markets, inciting revolutions, enabling criminal activities, generating media stars, facilitating commerce, reporting on news events, promoting friendships, and causing interpersonal conflicts. Yet an entire network can collapse overnight as people move on to better products and better communities. Such is the nature of dynamically linking system structures.

Terrorist Organizations as an Example of Dynamically Linking Structures: Terrorist organizations can be viewed as social networks held together by a common violent purpose. Such networks have learned to leverage Internet-based applications for communications and organization. However, a terrorist network can use any means of communications to include carrier pigeons. Like other social networks, terrorists join and form local groups within the dynamically linking system. Unlike other social networks, however, leaving the terrorist organization can be quite difficult, as it will be viewed as a betrayal that requires violent response. The ability of terrorist groups to take action without clear central authority makes this system difficult to stop. The ability of participants to secretly hide within the general population of society further makes this system difficult to find. If one terrorist is captured or killed, multiple new recruits might fill the gap. If a terrorist group/cell is broken apart, other groups might extend links to absorb surviving members. If individual terrorists or groups are captured, links can be quickly broken to protect the total system. Essentially, links between terrorists are constantly adapting to support the survival of the organization and the achievement of violent actions. As the only hard rule for associations is a commonality of violent purpose, terrorist systems are exceedingly difficult to destroy. One can attack the commonality of purpose or the integrity of the system parts. If all terrorists start to have doubts about the organization, then the system will weaken. If one terrorist is not sure whether another terrorist is a traitor, then the system will weaken. Otherwise, this dangerous system can stretch out to all the societies of the world and hide in the most obscure places.

Crisis Response Activities as an Example of Dynamically Linking Structures: My last example for dynamic linking is the system of first responders, military forces, government officials, and the general population all work together to respond to a crisis event. Under great threat or disastrous outcomes, the normal processes of society will stop. Then, the survivors will try to reorganize into a dynamic system to increase the odds of continued survival. If attempts at reorganize through new dynamic links fail, the people might either disperse or engage in conflict for resources and shelter. Advance planning and the training of people for roles during and after crisis can increase the ability of the survivors to form a new social structure. New associations need to be formed based on the evolving conditions of a crisis. For example:

- How should the police and fire department coordinate against each type of crisis?
- Under what conditions should military forces be called into action?
- Who in the government has relevant responsibilities?
- Where should people go as the crisis event changes?

These are all decisions that can be supported through studying representative crisis scenarios and determining potential courses of action. With knowledge of an appropriate step-by-step response, the people as parts of the new dynamically linked system can focus on precisely forming and breaking links. Breaking links is necessary at times when the capabilities of the parts are limited and/or the needs of the parts have different priorities. First responders must focus on those in greatest need. Government officials cannot take all phone calls. The people must sometimes focus just on personal safety. Therefore, the societal structure during and after a crisis is delicate. However, being able to reform a structure is incredibly important, as the imperfect structure is the foundation for building the future.

2.3.4 Dynamically Influencing Structures

Finally, some systems do not have any kind of lasting structure at all. This means that the parts do not have any permit or short-term dependencies on one another, and enduring processes cannot stretch across the parts. However, the parts still work together through interactions. Momentary associations that do not form links can still pass force, energy, substance, and/or information from one part to another. What makes these parts a system is then either some commonality of purpose among the parts that synchronizes with interactions or some connection of the parts to a common reference frame for basing behaviors. Through the constant interaction of parts, we can then say that the structure is completely dynamic and based on the parts influencing one another.

Parts influencing one another do not have the mutual pull to keep systems from breaking apart. Therefore, the parts have to independently anchor themselves to the spatial region of the system even as they reorient, move, and change their composition. In some cases, the environment for the parts will provide a boundary in which systems behavior can form. In other cases, the parts have independent controls that keep them within the proximity of one another for interactions.

Interactions among parts in a system can pass information that allows parts to coordinate behaviors, substances that allow for operations and proliferation, and energy that sustains operations. One part can also push or drag another part with a force appropriate for the reference frame of the system to get the other part better aligned with system behaviors. For example, if one part hits the boundaries of the system's environment, it can translate that force toward hitting other parts to prevent all the parts from crowding along the boundaries.

Thus, despite the lack of permanent and temporary links, influence-driven system structures can still reach stable states of dynamics. However, instability can emerge with increasing number of parts, intensity of movement among the parts, concentration of parts in one area, and deficiency of part interactions in other areas.

Virus Growth as an Example of Dynamically Influencing Structures: The effectiveness of viruses is that they travel far in the body and from the body. An airborne virus in the environment is potentially more contagious than one transmitted by fluid contact. Viruses that survive longer in the environment and viruses that infect multiple animal species are also potentially more contagious. In the body, the objective of all viruses is to invade cells and use the cellular components to replicate countless more viruses. As viruses are DNA or RNA strands surrounded by proteins, they are not fully formed life in the external environment. Therefore, viruses should not have any ability to form associations with one another to create a dynamic system. However, once a virus has invaded a cell, the cell components then become a part of the virus lifecycle. At that point, systems thinking would pose the question of whether associations can emerge. We would still expect these associations to be links between viruses, but can viruses in the cell control activities such as protein generation to influence the behaviors of other viruses? Research in this case is ahead of system thinking because the study of propagation rates and patterns in the vaccinia virus infecting monkey liver cells conducted by Geoffrey Smith's team shows that the viruses bypass cells that have already been infected by other member of their group [36]. Thus, the vaccinia virus is spreading faster than initial predictions. Apparently, the virus protein structure recognizes changes in surface protein characteristics for a cell infected by their fellow virus. This is, therefore, a classic example of a dynamically influencing system structure.

White Blood Cell Activities as an Example of Dynamically Influencing Structures: In the body of animals, the purpose of white blood cells is to identify, destroy, and consume enemy invaders and the infected cells of the body. Of the white blood cell types, the B type lymphocytes produce antibodies that identify abnormal entities, the T type lymphocytes recognize the abnormal entities and kill them, and the phagocytes to include the large macrophages finally consume the abnormal entities. Like viruses in the body, the white blood cells travel far and fast along the blood stream to go to points of infection. Therefore, they influence one another through the execution of their individual functions. Yet, there are no direct links between the cells as they can be far apart. The totality of dynamic influences, based largely on antibody signals, is the body's immune system. The power of this system is that all the parts are anchored to the reference frame of the body and guided by the conditions/signals of the body. As long as the body is viable and enables the production of more white blood cells, the system will fight on.

Nanotechnology as an Example of Dynamically Influencing Structures: Nanometer-size man-made molecular structures are categorized as nanotechnology. The potential of these structures in performing mechanical functions in a variety of environments to include the human body is being explored in scientific research [37]. The size of nano-devices does not permit elaborate communications capability to form long-range links. However, these devices can integrate into a fixed system construct through contact and they can cluster to create material properties. A more effective use of nano-devices could rely upon their ability to influence one another through contact and spread out across the environment like white blood cells. Nano devices can perform different operations, such as changing an organic molecule in the body or an inorganic molecule in the environment. Each type of device can build upon the accomplishments of other nano-devices. In this manner, nano-devices can, in theory, self-replicate if the raw materials for building themselves are available in the environment. By mimicking the behaviors of organic systems, systems based on nanotechnology parts can grow to immense complexity.

Democratic Political Process as an Example of Dynamically Influencing Structures: The process of electing leaders in human society is a dynamically influencing system because each eligible voter in the society must be given the freedom to make a personal choice. Direct links between candidates and voters, such as the buying of votes, and direct links between voters, such as pre-election day commitment of votes within groups, are strictly prohibited. Therefore, the only associations that can officially exist between the parts/voters are that of influence. If the parts are polarized and committed to candidate positions, then the system is not dynamic and the influence is not effective. The outcome of elections is, thus, dependent on the size and commitment of each side. If there are parts that are open to arguments and new ideas, then the system can be dynamic with influence pushing voters from one candidate to another. For this type of dynamically influencing system, the number of parts is limited by eligible participants, the range of associations is governed by laws, and the purpose of the system is to satisfy a defined process. The complexity in this system resides in ways to influence the possible outcomes of elections. One approach for studying these associations is through statistics. However, we can also try to understand the specific associations through uncertain reasoning techniques, as discussed in the previous section. Is it possible to understand the causes and effects on individual system parts? And is it possible to control the proliferation of effects across system parts?

In the first two sections, I focused on ways to find and study the parts and associations that go into systems. In this section, I focused on showing the breadth of system structures that can be formed by the parts and associations. The generalized examples presented should not be mistaken for the specificity of real-world systems. When faced with a real-world system problem, we can start with finding the parts and associations to formulate the structure and system dynamics. Alternatively, we can start with a general understanding of system structural type and then seek to discover the behaviors of the specific parts and associations. Neither of these paths will be simple if the system has poorly understood boundaries, intense interactions with the environment or other systems, and/or dramatic shifts in its form. This then leads us to the next three sections on the qualities of complex systems and the final section that summarizes how systems form through integration.

2.4 Boundaries: How to Define the System

The first statement about system boundaries should be that all boundaries are artificial constructs to make the study of systems more manageable. As a result of boundaries, the total system that is the universe is broken down into chunks within chunks so that the capabilities of man can measure, study, and perhaps control.

Each chunk can be regarded as a defined system or a subsystem of a greater system depending on one's perspective. Some chunks have obvious fixed boundaries, such as a mechanical device. However, one can still wonder whether the real system should include the person using the device as well as the entire operations the person is supporting with the device. Other chunks have boundaries that are expanding or contracting, absorbing or expelling parts, and spontaneously present and not present.

Although the structure of a system drives the nature of the boundaries, it is important to note that there does not have to be a direct correlation between the type of structure and the boundaries. For example, all four structural types discussed in the last section can be within a fixed boundary. Even when the system parts are only dynamically influencing one another, the range of dynamics can be within a boundary that is rigidly defined. Parts hitting the boundary, such as the surface of the body that contains white blood cells, might literally be prevented from escaping.

The other point to note before we jump into boundary types is that there is a difference between systems to which establishing boundaries is impractical and systems to which we have not yet figured out the boundaries. Many complex systems have boundaries that are difficult to determine, but this does not mean that they are truly unbounded systems. In our exploration, I will propose that unbounded is a specific boundary type that is diametrically opposite to a system with fixed boundaries. Between the bounded and unbounded system, there are other boundary types worth discussing.

In the effort to understand the boundaries and behaviors of systems, this section will further explore the utility of different research philosophies and associated methodologies. We will see how the definition of boundaries can change our perspective on system dynamics. And we will see how the definition of boundaries often depends on the reference frame in which we want to study the system. A system in one reference frame can have a completely different boundary type than the same system in another reference frame. Thus, we might care about one reference frame for control and another reference for understanding collateral effects.

The first type of boundaries, as shown in Fig. 2.26, is a boundary that sets a maximum range for the system structure by using a measuring point within a reference frame for system parts and associations. Thus, this boundary is fixed and the parts are bounded relative to the reference frame. The whole system along with the measuring point can move in the reference frame, but the boundary will remain the same as long as the ranges from the measuring point stay the same. If the system has a fixed structural configuration, then the boundary is simply the outer most parts of the rigid structure. Depending on the reference frame, the parts along the boundary might be the only parts within the system that interact with the environment or another system. This determination of which parts are interacting across the boundary with elements external to the system is perhaps more complex and important for systems with clustered, dynamically linking, and dynamically influencing structures. In the case of dynamically linking or influencing system structures, the level and nature of dynamics as well as the commonality of characteristics between parts may be what determines whether associations are within

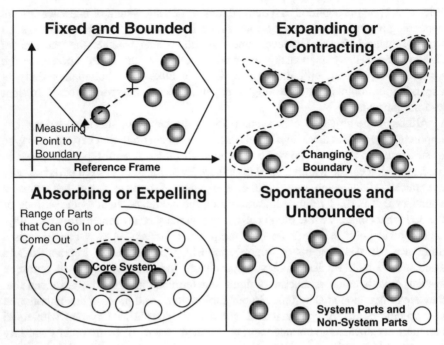

Fig. 2.26 Types of system boundaries

the system boundary or extending past the system boundary to parts that are not in the system. If a part is in the system, then a fixed boundary will prevent the part from moving beyond the maximum range. In some cases, the boundary is an actual barrier for containing the system, and the barrier might be a part of the environment, such as an ecological system being contained by the boundary of a lake. In other cases, the barrier is simply a behavioral rule of the parts where the parts will self-limit their dynamics at a specific range.

The second type of boundaries is a boundary that expands or contracts because system parts are pushing or pulling the boundary past fixed states. This expansion or contraction might only change the system's shape in a reference frame, or the volume might also be changed to create more distance between parts. The fluidity of the boundary and the level of resistance against initial change are system characteristics. Clearly, fixed and firm system structures will promote fixed boundaries that resist expansion and contraction. If an expansion or contraction is forced upon a fixed structure, then the system will typically want to settle down to another fixed structural and boundary state. For other types of system structures, a fluidic boundary can serve a variety of purposes. In a clustered and morphing system, the boundary might reflect how the system will engulf other entities in the environment or move around obstacles. In a dynamically linking system, the boundary might reveal how the system will expand to gain dominance. In a dynamically influencing system, the boundary might show the impact of increasing or decreasing dynamics. Sometimes, measurements and research will initially reveal more about the behaviors of the boundary than the structure of the system.

The third type of boundaries is a boundary that is absorbing and/or expelling system parts. This boundary results from systems whose structure is in a state of transformation or flux. All four types of system structures previously discussed can shed parts or take in new parts to change their structural configuration. Depending on the nature of the parts and associations as well as the structure of the system, parts can be removed or added to the system at specific rates and to specific sections. The porous boundary can, therefore, change shape as the system changes or maintain its shape to force the system into increasing or decreasing its density. The rules regarding which parts are going in and come out of the porous boundary are characteristics of the system. Parts can be pulled in or pushed out by existing links with the system structure. Parts can force their way in or out by creating links with the system structure. Further, parts outside the structure and parts inside the structure can have natural pairings that promote integration.

The forth and last type of boundaries is a boundary that is spontaneous and temporary. Parts can be associated with the system or outside the system at any time depending on dynamic conditions. As the system is essentially unbounded, this type of boundary is often not effective at describing system constraints. However, the boundary can describe system states. There may not even be a process for parts to cross the boundary in joining or leaving the system, as the unbounded system may not have a core set of parts. All four system structural types can be unbounded if the structures are prone to collapse and reorganization. However, dynamically influencing structures are more suited for unbounded characteristics. For example in the case of virus propagation, one boundary could be the entire environment of an infected body. However, the virus is designed to spread from body to body as well as from species to species in some cases. So other boundaries could be clusters of outbreaks, regions of epidemics, and national borders. The viral propagation system in this perspective can be viewed as an unbounded system where the spontaneous boundaries are our constructs to contain the virus. The virus then mutates in attempts at breaking past these boundaries.

The artificial definitions of boundary types above are established to help us understand the nature of system boundaries. Unfortunately, the actual determination of boundaries for real world systems can be challenging as systems tend to not fit cleanly into bins. There are two interrelated strategies for a real-world boundary definition. First, we can establish boundaries to match our objectives and methods for controlling the system. For example, we break down national voting in the democratic system of the United States by states, congressional districts, and precincts. Second, we can discover natural points where boundaries can be defined by studying the behaviors of the system. For example, a bacteria growth system might be naturally constrained by the extent of a nutrient rich growth environment. The results of the first strategy should not conflict with natural boundaries, and the results of the second strategy should also help us with control. The challenges come in when neither of the two strategies work or when system behaviors extend into other reference frames to yield collateral effects and hidden consequences. When boundary definition strategies do not work, it generally means that the controls are not effectively controlling the system or that the natural divisions are not really clear and lasting divisions. When system behaviors in a different reference frame are yielding dramatic consequences, it generally means that the initial dimensions in measuring the system were not broad enough. Parameters thought to be irrelevant actually have impact or there are unknown parameters as well as parametric ranges in the problem.

To wrestle with real-world systems with unclear boundaries, researchers have at times declared such systems initially as all unbounded to facilitate unconstrained searches for system parts and associations using techniques previously discussed in Sects. 2.1 and 2.2. For example, agent-based models can test the dynamic ranges of system parts to see where unknown parts to the system are hiding in the real world. This search, however, can become overwhelming when dealing with systems, such as human society, with millions and billions of parts. Thus, other approaches for assessing the system from a macro perspective have been adopted by researchers in different fields. Though each field will advocate for its specialized methodologies, we can place these methodologies in a general systems framework to understand their value and limitations. Three dominant methodologies are discussed below.

2.4.1 Methodology of Iterating a Conceptual System Model

As noted earlier, Soft Systems Methodology (SSM) seeks to iterate a conceptual model of the system until enough insight and accuracy are gained to affect the system. The general tenet of this methodology is that no system model can be an exact replica of the real-world system and that the act of measuring the real system changes it. Instead, the model is an abstraction with compromises made due to inability to measure the real system beyond a set point, record all possible data associated with the real system, identify all the parts of the real system, and/or avoid disturbing the real system. For example, if one wants to perfectly model a clock, one will have to record the structure of the clock down to the atomic and subatomic levels and identify all the effects on clock from the friction of the gears to the transfer of heat from the environment. One also cannot take apart the clock or even touch it in any way while measuring it. Giving up on perfect, we can create a conceptual model of the clock with the shape of all the gears to show how the clock works. Therefore, the objective of SSM is to create and leverage an effective conceptual model.

The SSM iterative cycle described by Peter Checkland is practical and problem-resolution focused for the business community. We can describe his cycle in a slightly more general perspective and with less advocacy to be one of multiple ways to wrestle with the unclear boundaries and behaviors of real-world systems. The iterative process, as shown in Fig. 2.27, begins with recognizing a complex real-world system. There are existing ways to measure the parts and associations in the system, but these ways might not be able to capture all the structure, dynamics,

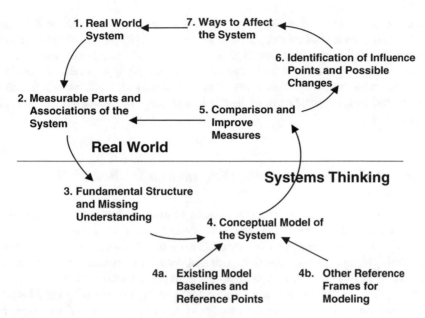

Fig. 2.27 General cycle of soft systems methodology

and boundaries. One can have missing understanding in a reference frame, and one can have completely missing reference frames. From what the real world can provide, systems thinking then projects the structural characteristics of the system and the missing understanding between the measurable and complete reality. The inconsistencies help us to create an initial conceptual model. Based on the system structure, one might be able to find other systems with similar structures to help build the conceptual model through comparative analysis. Similar behaviors in other reference frames can also help define the conceptual model for resolving inconsistencies.

The initial conceptual model can be brought into the real world and compared with existing measurements. This comparison might enable the improvement of ways to measure the system, which then leads to further refinement of the conceptual model. After iteration, the conceptual model might reveal points for influencing the system and ranges of possible change within the system. Using these influence points, we then want to find ways to affect the systems in an understandable manner. This process of interacting with the system will probably not be initially precise. However, one might be able to get to an acceptable level of control through additional measures and additional refinement of the concept model.

We do not have to follow SSM rigorously to apply the method of iterating and testing an imperfect model to understand the hidden nature of a system. In fact, the model can be abstract when there is limited data about the system, and the model can still have value if new ways to measure the system can be gleamed from studies. The weakness of an iterative methodology is that conceptual models that can be easily iterated often cannot be extremely dense in parts and associations. Further, the philosophy that one cannot exactly model reality may cause us to prematurely give up on the idea of more detailed modeling approaches. We need to, therefore, know when to break away from this philosophy and switch to other system modeling approaches once iterative conceptual models have served their purpose.

2.4.2 Methodology of Eliminating Intermixed Boundaries

The alternative to using the temporal domain to enhance systems understanding, such as through iterations, is the idea of studying systems with multiple potential boundaries by eliminating sets of boundaries or making some boundaries more important. This spatial simplification methodology is most commonly found in the study of macro human interactions across the world, as shown in Fig. 2.28. For centuries, the most obvious boundaries around people are that of political states represented by territories controlled through systems of government. However, as

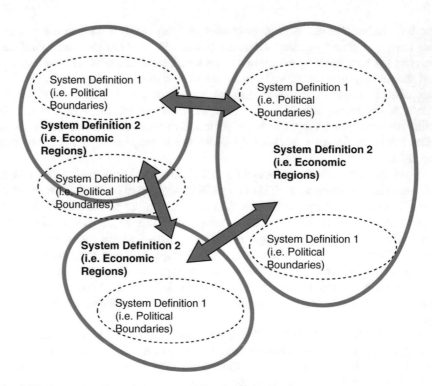

Fig. 2.28 One type of boundaries more dominant than another type

people migrate, companies trade, and banks move currencies across state borders, scholars have wondered whether there are other boundaries around people that are more definitive of systems changing the world. Before discussing this second set of population boundaries, let us first explore the idea of state boundaries and how such boundaries have already displaced older boundaries based on human associations.

From a systems perspective, the nation state is an artificial construct of man to sustain the power of government and the ability to establish militaries for enforcing and expanding state boundaries. The natural parts in state systems are people. The natural associations across the parts that can support or hinder the strength of government are the common ethnicity between people, the shared cultural experience among the people, the people's embrace of prevailing ideology, and/or the people's dependence on an integrated economic system. When state boundaries are aligned with the boundaries of these natural associations, the government tends find a form of stability. When state boundaries divide ethnic, cultural, and ideological boundaries, as in parts of Europe, the Middle East, and Africa, tensions within state boundaries can become very high. Countless political science papers have been written about the tensions within states and the tensions between states due to misaligned boundaries between the government and common human associations. A part of the discussion is how governmental boundaries have been changed through civil war and invasions to achieve better alignment.

The boundaries of human associations can also influence forms of government as well as conflict behaviors. Government boundaries tightly wrapped around ethnic boundaries can promote racist policies. Government boundaries tightly based on strong culture can turn nationalism into a fascist policy of cultural expansionism. Alternatively, many governments are enforced by ideologies and promulgate their ideologies. One of the earliest ideologies for governing is that of a ruling class (nobility) running the government and the head of state (monarch) coming from nobility. A variation of the ruling class system is the ideology of the head of state coming from the religious caste as a member of the clergy. And, more recent in history, is the ideology of representative democracy where the head of state and government officials are elected from the general population. Along this view of government and human associations, there are two ideologies that have not been popularly embraced across human associations but have, at times, been successfully promulgated by the power of government. First is the idea that one single person should have absolute power over the people. Although this idea tends to be rejected by the people, totalitarian governments have formed based on the head of state's grasp of military power. Second is the idea that only the wealthy should have power over the people. Though disliked by the majority of the people without wealth, governments owned by the wealthy have formed based on the politicians' allegiance to money. These last two examples are important from a systems perspective because they show that the boundaries of government artificially established by man can overcome other boundaries among people to be powerful and enduring.

Returning to the methodology of eliminating types of boundaries to focus on the dominant levels of system interactions, the nation state appears to be one logical set of boundaries, as we can treat ethnicity, culture, and ideology as merely forces with the boundaries of states. The interactions between nation states are then governed by political decisions and explored through political science theories. Though greatly over simplified, we can say the *Theory of Anarchy* argues that states brutally compete with one another for survival [38]. The *Theory of Realism* argues that the decision to wage wars must be tempered by rationality [39]. The *Theory of Neorealism* argues for the overcoming of hostilities between states [40]. The *Theory of Neoliberalism* argues that states can achieve limited cooperation [41]. And *Theory of Liberalism* argues that states can collaborate as partners in the global community [42].

Beyond the theories regarding motivations for political decisions, there are theories that argue that political decisions are merely reflections of the needs of system structures within states and between states. The *Organic Theory* argues that states behave like predatory living entities and must seek to devour more territory to survive [43]. The *Theory of Constructivism* argues that states may have an evolving self-identity that governs behavior [44]. The *Theory of Democratic Peace* argues that democratic states are motivated to not wage war against one another, as the people pay the price of wars [45]. And the *Theory of Institutionalism* argues that global financial institutions and corporations are shaping the actions of states [46]. The challenge with all these theories is that, as long as the system is defined by political boundaries, the interactions between the systems are always vulnerable to the irrational and emotional decisions of leaders. At times, psychological profiles of individual world leaders may provide more insight than studying system dynamics.

The dependence of state systems on leadership decisions leads us to question whether leadership decisions are really what shape the dynamics across the global human population. If we eliminate state boundaries, what is the second set of boundaries between people for defining systems? One type of new boundaries is based on global geography. Are there advantages accorded to people of an entire region, which may include many states, due to the region's geography? In response, the *Heartland Theory* argues that the people at the center of mass in a continent will have advantages in territorial reach [47]. So whoever can dominate the heartlands of Asia and Europe will dominate the world. In contrast, the *Rimland Theory* argues that the regions with both sea and land access have the advantage in global reach [48]. So whoever can dominate the rimlands will dominate the world. We can draw boundaries around heartland and rimland geographical areas. However, the *World Systems Theory* makes the argument that a region's economic status is what drives its success in global dynamics [49]. Boundaries must, therefore, be drawn around core regions with advanced technologies, diversified economies, and educated workforce. The core system, according to theory, is then surrounded by buffer semi-periphery regions with industrializing economies and growing level of skilled labor. Finally, the periphery regions are beyond the boundaries of the buffer regions and have nonskilled labor, weak governments, and primarily exploitable raw materials.

Related to world systems are the *Theory of Modernism*, which argues that regions or states advance incrementally to the level of modern society [50], and the *Theory of Dependency*, which argues that regions or states in the periphery are trapped by their dependence on the core [51]. In all these regional theories, if we eliminate state boundaries from consideration and focus on the regions, then we can suggest that all the countries in a region will share similar system structures. Regardless of theories, we should expect to see similar system behaviors and issues in one region and different system behaviors and issues for another region. The debate, however, is whether these boundaries are more dominant than decisions of the state.

In discussing so many theories, I know better than to be caught in passionate debates against advocates. I present the theories to explain an overarching methodology, which may not be recognized by all the scholars advocating theories. That is fine. In order for the methodology of eliminating boundaries to work, advocates of specific boundary types must merely argue that their definitions are the most dominant. This is despite the reality that all the boundaries might be interrelated and all the theories might have conditions where they are not fully relevant. If we are to drop a set of boundaries, then the argument about dominance has to be made and compromises have to be accepted.

Lastly, this methodology does not have to be restricted to human population systems. Almost any group of associating parts that can be defined by multiple types of system boundaries can go through the assessment of which boundary definitions are more important. The key is to not to lose sight of reality, recognize that dropping boundaries is only for the convenience of studies, and understand that all boundaries impact the system.

2.4.3 Methodology of Summarizing Lower Order Systems

For systems that are pervasive in their spatial reach, the boundaries that we care about are the range of fidelity for measuring internal system dynamics and the range of time intervals for taking measurements. The methodology for leveraging fidelity and time interval is to find ways to roll up vast amounts of discrete lower order data into macro-level parameters that describe the pervasive system or to find ways to directly collect summary data based on the macro-level parameters. The economy is an example of a pervasive system. Earlier, I have noted economics as one of four main structures connecting the human population, and I have presented other scholars' theories on dividing up the world into different interlocked economic regions. The global economy can, however, also be viewed as a single system with the research methodology focused on what macro-dynamic parameters and models accurately represent the interactions and outcomes of countless firms, distributors, and buyers, as shown in Fig. 2.29. The range of these macro-dynamic parameters then represents the boundaries of the global system.

To demonstrate the methodology of macro-dynamic analysis, the most important parameter in macroeconomics as recognized by Adam Smith is the capabilities and capacity of the people in producing goods and services of value [52]. Thomas Malthus then recognized that changes in the size of the population impacts economic activities [53], and Ricardo [54] recognized that the ability to specialize and produce in large scale creates a comparative advantage. Marshall [55] further

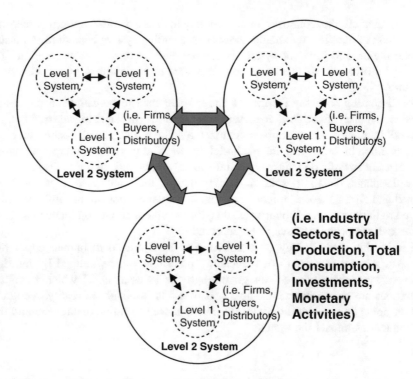

Fig. 2.29 Macro-system based on summation of lower order systems

explained that total production of specific types of goods and services and total demand for those goods and services are interdependent and have dynamic cycles, as long as firms have the freedom to produce and people have the freedom to buy. The dynamic cycles in the economy can be further understood by measuring time for market awareness, time for production, and time for deliveries [56]. The importance of information about market needs and the logistics to match changing market needs are further explained by Von Hayek [57].

Decades of scholarship by economists across the world have led to parameters and processes for measuring the macroeconomic system without having to model all the parts and associations in the economy. Specifically, these parameters take aggregate inputs and outputs from the totality of industry sectors, market segments, buyer population groups, information channels, and distribution networks. The challenges of summarizing the dynamics of lower order systems to bind the macro-system are: (1) whether the range of measured data over decades is enough to identify all the behaviors and potential behaviors of the macro-system; and (2) whether the economy can be effectively influenced through these macro-parameters. These challenges have led to some theoretical debates as further described.

The first debate is over the existence of longer cycles of change in the economic system [58], and the possibility that even the most advance regions of the system is still evolving in structure [59]. In fact, *Evolutionary Economics* has become a dedicated field of study to focus on the natural forces for changing capitalistic economies.

The second debate is over how and to what degree the government should influence the economy to achieve sustained growth. John Maynard Keynes argued that intervention policies are required by the government to reduce the impact of inflationary and depressionary forces in the economy [60]. In contrast, Milton Friedman argued that the economy is more self-correcting [61]. The resulting Chicago School (monetarists) of practice is, thus, focused on controlling the monetary supply as the primary means to stimulate the economy to self-correct.

The third debate is whether there are components of the economic system that are disproportionately concentrating wealth relative to their value. This surfaced during the industrial revolution and might surface again. Karl Marx described this as "surplus value," [62] but the resulting remedy of communism became a failed form of government. Nevertheless, such imbalances in the economy might still exist, such as the increasing concentration of wealth in the 1 % population of the United States as noted in media [63]. If so, are there any ways to correct possible inefficiencies without abandoning the entire economic structure?

As we consider the debates that still go on in macroeconomics and the periodic inability by scholars to project major economic events, we should remind ourselves that the approach of macroeconomics is just one method for dealing with complex systems. The boundaries of the economic system do not have to be drawn at the current levels. With advances in "Big Data," we might be able to model economic behaviors from the overall dynamics of the firms directly up to the behavior of the global system. How the actions of one CEO, the performance of one company, and the buyers of one product impact the global economy can be captured in the economic system models of tomorrow.

I close this section with the reminder that system boundaries and the definition of systems are established by those who study them. Some boundaries have been set by academic communities for so long that no one tends to question their limitations and effectiveness. However, we must always be clear about what is reality and what compromises we have made to study reality. When existing boundary definitions are hindering our studies, then new boundaries as well as new methods for establishing boundaries should be explored. The success in studying a system often depends greatly on how the study begins. Toward this purpose, I have only presented some methods for establishing boundaries.

2.5 Interactions: How the System Behaves

Together, we have explored the structure of systems and the boundaries for defining systems. I have further emphasized that systems exist because of their dynamics, the changes between parts and associations for the system, and the changes within parts and associations for the system. These changes can be further studied as a

Fig. 2.30 The purposes of
system interactions

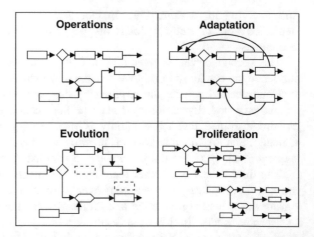

whole at the systems level assuming that the parts are working together [64]. If the
parts are not working together or even influencing one another in an integrated
manner, then we do not have a system. If the parts are having problems and
difficulties working together, then we are facing concerns about system failure to be
addressed in Chap. 3. For now, let us assume that we have working systems that are
able to interact with their internal parts, the environment, and other systems. Our
next step in exploration is to then look at the ways and proposes of interaction.
Similar to the decomposition endeavors of prior sections to find broad sets of
system characteristics, system interactions can be divided into four conceptual
types, as shown in Fig. 2.30. Unlike the structural and boundary types, which may
have unclear divisions but are nevertheless mutually exclusive, the types of system
interactions to be discussed can all emerge over the course of a system's life cycle.
Further, as we explore the types of interactions, it will be clear that one type of
interaction can impact the dynamics of another type of interaction to make system
behaviors very complex.

2.5.1   Operational Interactions

At the fundamental level, the purpose of all systems is to operate as defined by
allowing the parts and associations to work together. The operations of a system can
be captured in an overarching process or series of processes that describe what each
group or subgroup of parts and associations is doing. Processes exist even when the
associations between parts are changing as long as the dynamics between parts are
not random but serve the system's purpose. Processes also exist, even when the
parts are being added or removed from the system, as long as the changing of parts
across the system boundaries is acceptable based on the design of the process.
Finally, processes can be momentary in unbounded systems or systems with

incremental as well as sporadic associations. However, there needs to be ways to project when the system will undergo certain processes.

As few systems are perfect by design or through formation, system processes and decisions can contain degrees of variability with cases of lower performance and levels of imprecision leading to errors. Improving the consistency of performance, such as production rates, and avoiding errors that stem from performance, such as product defects, thus, have been matters of great research focus for human designed systems. This is in addition to the fundamental system design challenge of creating structures and dynamic processes that make achieving objective performance and behavioral results feasible.

Assuming that acceptable performance can be maintained and error rates can be kept at acceptable levels, we then have a functioning system that will operate with the purpose of sustaining itself, producing outputs, affecting the environment, and/or controlling other systems. The idea of a system merely sustaining itself seams pointless for human designed systems. However, for systems where the parts have value, such as people in a social system, the simple purpose of processes for enabling parts to live and thrive can be noble. Systems that formed in nature tend to follow one test of acceptable performance, which is survival under harsh environments. To pass nature's test, systems must act with enough consistency and precision to fit within the changing dynamics of the greater natural system. Slight mistakes in action could lead to death. Slight deficiencies in capability could lead to the extinction of an entire species. At the same time, systems in nature work together to help one another survive. Operations support one another and natural processes connect with one another.

Beyond a system acting just to keep itself viable, the processes can produce outputs such as products, raw materials, energy resources, and information resources. For systems in nature, their outputs, such as oxygen from plants, can help other systems to survive. For human designed systems, the outputs often define the purpose of the system. The most obvious outputs come from factory processes, and the most complex outputs might be the terabytes of information generated by software system processes connected across the World Wide Web.

The counterpart to a system contributing something new to the environment is the system performing actions that affect existing components of the environment. These actions include transportation, assembly, alteration, demolition, storage, and consumption. In fact, most systems consume from the environment and contribute to the environment in some way. The counterpart to a system contributing something new to another system is the system controlling or attempting to control another system. These controls include physical changes to the other system, forces upon the dynamics of the other system, instructions to the other system, and parametric inputs to the other system for basing dynamics. If the interactions between the systems are tight enough, then we can consider whether we are studying a system of systems or a system with subsystems.

Years of researching systems operations have led to a series of methodologies and techniques for managing the operations of human designed systems. One popular methodology, which focuses on identifying and improving core measures,

Fig. 2.31 TQM based view of operations management

has been Total Quality Management (TQM). TQM, referenced earlier and shown in Fig. 2.31, recognizes the importance of both leadership and participating workers in a modern corporate organization. At all levels, employees need to understand that they are producing in response to the needs of customers internal to the corporation so that processes connect. Further, the total production must respond to external corporate customers, which is the bottom line.

The involvement of leadership and workers in the processes of operations enables the system itself to focus on operational measures to improve and sustain quality. Performance issues in satisfying processes should be continuously identified and resolved. Then, the members of the organization/corporation need to be trained or educated to prevent performance issues from reemerging. Through this empowerment of the system to achieve optimum operations, end products will, in theory, have better total quality. Continuous identification and prevention of component and assembled product errors will lead to the elimination of waste. TQM was popularized in American manufacturing in the late 1980s and early 1990s to reduce defective products. Later, as organizational processes became more agile with computer-driven design, production, and coordination, management focus appears to have shifted from metrics centric approaches to process capture and change based techniques such as Lean Six Sigma, which was also referenced earlier. Lean Six Sigma, for example, relies on external experts (Black Belts and Master Black Belts) to map the processes, isolate statistically verifiable performance issues (Six Sigma standard is over 99.99 % error free), design/test corrective actions, and improve the entire process flow (lean design of processes). Such techniques not only optimize operations but can lead to a transformation or evolution of the system to be discussed. Employees can be removed or added, business units can be changed, and processes can be completely replaced.

2.5.2   Adaptive Interactions

Beyond a system merely operating against a set of processes, the operations of the system can be adaptive. In other words, the processes allow for changes in rates, procedures, outputs, and interfaces depending on the feedback received from system performance results as well as interactions with the environment and other systems. I use the term adaption in this context to represent responsive changes in the way the system operates but not changes in the structure of the system. For example, human beings are remarkably adaptive systems compared to other animals, even though our bodies are rigid systems in a similar manner. What allows for human adaptiveness is largely the superior brain that receives sensor inputs, records past experiences, and learns to act based on past experiences to optimize the human condition. From the learning process comes the body of human knowledge built and shared across the centuries. From the desire to improve the human condition comes innovation, discovery, and construction to control the human environment. Over the centuries, the human condition has gone from hunting and gathering to survive to traveling into space and underneath the sea. We have discovered medicines to alter the performance of the human system, and we have invented devices to enhance the capabilities of the human system. However, it is only recently, with the mapping of the human genome and developing of DNA modification techniques, that man can change the structure of his own system. Before we explore evolution, however, let us first delve deeper into the purpose of adaption.

On one level, the purpose of adapting through system interactions is to improve the performance of the system. The performance of the system is then tied to the mission of the system. In the case of companies, for example, the general mission is to gain higher market share, increase profits, and grow intellectual property and other assets. The mission of an organic system is to survive, reproduce, improve quality of life, and satisfy emotional and/or intellectual needs. The mission of military systems is to defeat adversaries, defend against threats, capture and control territory, and deter potential aggression. The ability to adapt might be critical to system operations if the operational environment is highly unstable or unpredictable. This would prevent processes from being designed to rigidly interact with the environment and other systems in the environment. Instead, the processes must be designed to adapt to unpredicted events, new drivers, and shifts in environmental conditions. The human mind is ideally suited for this type of unbounded system interaction. Artificial intelligence research and learning systems based on cognition theories are starting to move computers toward higher levels of adaptive capability.

In seeking to control system adaptiveness, the modern military concept of tactical operations as captured in the Observe, Orient, Decide, and Act (OODA) Loop created by US Air Force Colonel John Boyd is a good reference frame [65]. Although the OODA Loop was designed for combat in highly dynamic and often unpredictable environments, it can be used to describe the cycle of system processes required to adapt to any performance conditions. This expanded use of the OODA Loop is shown in Fig. 2.32. Essentially, in order for a system to adapt, it must

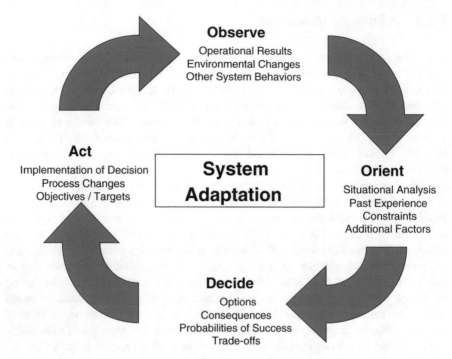

Fig. 2.32 Adaptive cycle based on military OODA loop concept

observe and establish an understanding of what to adapt against. This understanding can be achieved through feedback on operational results, changes in the environment, and the behaviors of other systems in the environment. For nonhuman systems or systems that enhance human capabilities, the sensors and metrics for enabling observations are the starting point of adaptation. Speed, cycle time, and fidelity of observations are then driving factors. With observations, the system analyzes the situation to determine a conceptual model, compares the model with past experiences consistent with Naturalistic Cognition Theory, and determines other factors as well as constraints that would affect decision-making.

The actual decision process would start with the adaptive options associated with the conceptual model of the situation, the understanding of consequences associated with each option, and the assessment of probability for success associated with each option. Before an adaptive action can be taken, the system must first decide on which option to take based on trade-offs. The challenge of the trade-off process is that few conditions have clear answers. In most cases, options carry negative consequences as well as risks of negative outcomes beyond just the potential benefits. The human mind is quite skilled at navigating through these opposing factors, but it is nearly impossible for rules-based computer programs to cope with the unpredictable aspects of operating environments that require adaptation. Thus, uncertain reasoning techniques and artificial intelligence is the path that must be taken.

The actual act of adapting is the last part of the response cycle, and all actions must have clear objectives and ways in which processes are adjusted to achieve the objectives. If the actions are perfect and complete, then the adaptation can stop, and the system might be able to achieve a new steady state level of operations. However, adaptation is often an iterative process where the system must interact with the shifting situation. Actions must be taken, and the outcomes must be observed so that the system can reorient further adaptive actions.

In an interaction involving two or more opposing combat systems, the OODA Loop is continuously executed on both sides. Often times, victory is dependent on the speed of this adaptive cycle and the accuracy of the actions. Getting within the cycle of the enemy can disable the enemy's ability to respond, as one is adapting faster than the enemy's ability to reorient and act.

Adaptive actions can be centrally controlled within the system or be taken individually by system parts. This depends on the structure and boundaries of the system and the ways in which the system interacts with the environment as well as other systems. Central control allows for the power of concentrated system-wide knowledge, situational awareness, and computation to be applied to actions. So the adaptive actions can be complex with deep understanding of total potentials and consequences. Adaptive actions by the parts can be faster in response to environmental situations, harder to counter because of many decision points, and highly complex as a result of mutual coordination. Even the human body has localized reflex actions that do not require commands from the brain. However, it is centrally commanded and locally responsive human systems that pose the choice of selecting the approach that is more adaptive. For example, some companies embrace the delegation of adaptive capability to highly trained frontline employees. Other companies want employees to rigorously follow the commands of leadership who will decide how to adapt. The right approach is probably situationally dependent, and the most adaptive company/system might be the one who can switch the control of adaptive actions based on needs.

2.5.3 Interactions to Evolve

As noted, all types of system structures and system boundaries, though not all systems, can be adaptive in their behaviors. Likewise, all types of system structures and boundaries can evolve. Evolution involves the changing of the parts and associations enabling system dynamics. This change can be within the structure and behaviors of parts and associations, and this change can involve the replacement of parts and associations. Evolution can be caused by external and internal forces, and changes might not always be advantageous to the purpose and mission of the system. Positive evolution is an adaptive event, but systems adapting through evolution might not be adaptive when they are not evolving. In contrast, adaptive systems might not be able to evolve. Negative evolution is a reflection of weaknesses within the system at the levels of parts, structures, and/or boundaries. Vulnerabilities to forces might cause the system to change to a state with less structural integrity and less ability to successfully operate.

Natural selection is a theory suggesting that, through random changes in the system and letting these evolved systems compete with one another in the environment, the most fit systems will survive in the end to dominate. Systems with poorly evolved features and capabilities will die off. The limitation of this theory is that it takes a step function view of how systems interact with the environment. In this view, what is best in the next step is given priority over what might be better for the future. If systems are intentionally evolved through reorganization to surpass previous systems, then the designers might wish to look at the changing environment and needs across the extended future, not just a series of next steps. One clear risk of a next-step evolutionary approach is reaching dead ends, such as mass extinct events as presented by paleontologists.

The key parameters in system evolution are the rates of change and the levels of change. For systems such as those based on information technology, the rate of technology advances has often driven systems evolution as well as the invention of completely new systems. However, whenever such changes occur, they create learning and adaptation changes for the human users. Therefore, product release cycles and product adoptions have been a major area of study in market research. Replacing systems too quickly or with too few innovations could lead to reduced profitability from prior system investments, user delays in adoption, and/or user perception of lacking value. Replacing systems too slowly or with too major a change could lead to reduced relevance for the existing system that should have been replaced, overwhelming of users, and/or loss of user willingness to change. One can argue that evolution is an inherently high risk but perhaps necessary level of system interaction. When pushing for evolution, both in human systems and human created systems, the process of transition from the current system to the next should perhaps be regarded as equally important as the structures and functions of the next system.

A method of transition is evolving systems in the course of reproduction or replacement. Changes in parts, structures, and boundaries for organic systems and systems mimicking organic characteristics can occur during mitosis and sexual reproduction. Changes in other systems can occur as a part of constructing replacement systems. Some systems can, however, evolve without being replaced, and some types of structures and boundaries are more conducive to changes while the system still operates.

The control of human organizational systems to evolve while continuing operations has been a matter of great research emphasis in management studies. The Fifth Discipline presented by Peter Senge and referenced earlier specifically argued for system thinking in enabling organizational change, and I have already noted that Lean Six Sigma offers techniques for changing the system as a part of changing processes. Because of the complexity in this endeavor and the ongoing research into controlling organizational transformation, I am not going to present any methodological diagrams. For example, the Balance Scorecard, as referenced earlier, argues for change as a cycle of setting Vision and Strategy, Communicating and Linking Performance, Planning and Target Setting, and Strategic Feedback and Learning. However, I do not want to trap anyone in formal steps given the complexity and diversity of system dynamics. A methodology might work for one organizational condition and not the next. So I will present some summary observations already made by other systems and management science researchers.

- Organizational systems are often inherently self-adaptive and self-organization based on organic behavioral properties
- Even with performance issues, a stable system will often resist change, and the resistance will increase with external attempts to control change
- Unstable systems will change in unpredictable ways. Thus, causing change could be easy, but controlling change might be difficult
- Complex systems will change in complex ways with possible untended consequences
- There may be hidden patterns and unforeseen forces that affect one's attempt to control change
- The right control points are, therefore, difficult to find
- The extended effects from a control point can emerge anywhere in the system at any time.

Thus, controlling the evolution of a human organizational system could be a myth. Nevertheless, managers and operations researchers might be able to nudge system changes in better directions. One can redesign to obtain the best processes and hire the best people, but the enduring culture of human organizations might get in the way. Gentle use of control points can yield significant results, and iterative use of control points might hone in on more preferred outcomes. So, when in doubt, systems evolution should perhaps be in small increments.

2.5.4 Interactions to Proliferate

Evolution is a state-to-state type of system transformation. A system can also be designed to dominate through proliferation instead of just through operations, evolved operations, and adaptation. The proliferation that allows a system to dominate the environment is growth in number of system parts and/or high rate of reproduction to yield greater number of systems. Proliferation is, therefore, another form of system interaction. In growth, the system structure or parts of the structure can be replicated to expand the system. The system can also have processes for building structures in evolved ways as it grows. In reproduction, the design of the parent system or integrated design of multiple parents is used to create offsprings. A high number of resulting systems is achieved through having many offsprings in one generation and/or having many succeeding generations.

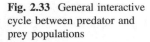
Fig. 2.33 General interactive
cycle between predator and
prey populations

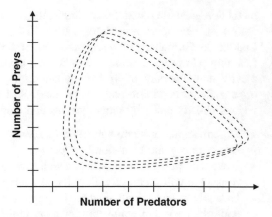

    System proliferation in environments that seek to achieve balance can be an
internally or externally regulated interaction. In internal regulation, system growth
and reproduction will slow down based on indicators that the environment will not
be able to sustain the size and/or population of the system. For example, food and
other resources could be dwindling, competition for resources could be intensify-
ing, and environmental conditions could be getting harsher to warn the systems to
adjust interactive behaviors. For systems that are not good at internal regulation,
such as poorly organized packs of animals, external forces or systems could help to
achieve balance. The most well-known proliferation balancing interaction is perhaps the predator and prey model as shown in Fig. 2.33. In this simple interaction, a
population of predator animals (such as lions) and a population of prey animals
(such as zebras) provide a check and balance in population growth [66]. As the prey
population increases to dominate the land, the predatory population will increase to
eventually dominate the prey. As the prey population is brought down by the
increasing number of predators, the reduced food supply for predators will cause a
lagging reduction of population. Once the predator numbers drop blow certain
levels, the prey population can grow again in a system-to-system coupled manner.
    The cyclical approach to achieving balance in potentially unstable systems can
be seen in a variety of system control schemes. For example, when using opposing
thrusters to stabilize a spacecraft in a low gravity atmosphere-free environment, one
often can only get the stability down to a small wobble called a limit cycle, as the
exact force to fix the system's position is not achievable. Limit cycles are common
to oscillatory systems when the oscillations are not growing out of control or dying
down. When the system is governed by more than two opposing forces, stability
and control are more complex. Control science is the discipline of identifying all the
dynamic metrics governing a fairly structured and bounded system (such as an
aircraft), determining how the metrics associated with one another (typically
through partial differential equations), and solving these equations often through
computer-based methods to gain control of the system. This discipline does
encounter difficulties with systems that are less structured and less bounded.

In environments with less balancing forces, unconstrained system proliferation has, at times, led to massive systems die-off. Bacteria cultures in the lab have demonstrated this behavior where mass growth has reached a point of nonsustainability. Then, suddenly, the population will die off to a level where survival in the environment is again stable. It is still uncertain whether human social systems can internally regulate growth. One can argue that competition for resources and wars between regions in the past has provided some constraints. With a global economy and rapid growth across many regions of the world, how to sustain the system has become a paramount question in the twenty-first century.

Real-world systems can be extremely complex in structure and undefined in boundaries. Therefore, the results of system interactions can be filled with unknown dynamics and effects. I do not want to trivialize the challenges in understanding system interactions in any way, for the book of management history is filled with failed attempts to shape the interactions of human systems. Even in war, with decisive outcomes during most tactical engagements, the consequences of the conflict may not be fully realized for generations. We might create new enemies in the course vanquishing the old. We might adversely change our system/our way of life while trying to adapt to threats and a changing world.

2.6   Quality: Measuring the System

Much of what I have discussed is focused on how difficult it is to fully and accurately define systems. In fact, some have argued that real-world systems can never be fully defined but only understood to a level where we can use influence to obtain specific results. The idea of establishing system-wide quality measures is, therefore, almost a false belief, as the systems will surely stretch beyond these measures. Nevertheless, it is not always possible for us to describe a system by all the parts, associations, and dynamic effects. If we need to make quick decisions regarding systems, it is sometimes practical to simplify our view of complexity, make compromises, and establish quality measures.

Quality measures born from compromises can be helpful in unraveling system characteristics. This is because the measures permit comparative analysis between systems when details are not well understood or when details actually distract us from common characteristics. Despite the differences between systems, common patterns often appear at the macro-dynamic level. For example, some human group behaviors might resemble the patterns of colony insects, and some Internet traffic patterns might resemble the flow of people and goods between cities. In such comparisons, the details of one system might help us discover the details of another system with less understood structures, boundaries, and interactions.

Now returning to the thought of making decisions about systems based on quality measures, reality sometimes calls upon us to choose which systems to monitor for potential failures, track as threats, leverage for opportunities, modify to expand capabilities, mimic in design, and consider terminating. The world is filled with systems and different ways to define systems, so many that we cannot model the world. Therefore, we must choose what to focus on and how to focus. Please note that I, at no point, advocate taking drastic actions based solely on quantity metrics without further studying the systems to be impacted. However, the time needed for studies may not be available during crisis situations. Making the perfect decision at the wrong time is pointless. Thus, I will try to cover a broad enough set of measures to enable crisis decision-making.

To facilitate decision-making, quality measures can be divided into three categories, as shown in Fig. 2.34. The first category is the qualities that enable a system to sustain its intended purpose, which includes the method of operations and objective results. The second category is the qualities that enable a system to compete in order to fulfill its intended purpose. This competition can be in a rules-governed environment, such as the commercial marketplace, or this competition can be an all-out conflict for survival, such as in total war. Finally, the third category is the qualities that enable a system to improve and move beyond its current state. While not all systems must be able to improve, systems that cannot at least improve through replacements and successive generations will most likely become extinct. Systems that endure beyond the need for improvements will often become obsolete.

There may be other ways to categorize or bin together quality measures, and my list of quality measures might not be complete or even properly named. Achieving a

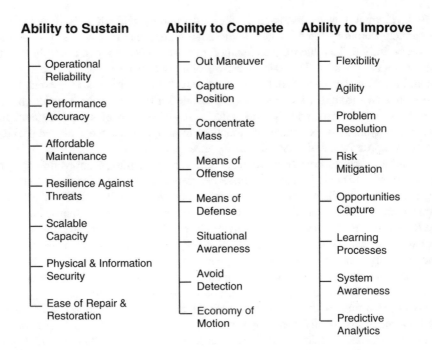

Fig. 2.34 Proposed categories and measures for quality

commonly agreed to taxonomy for measurements has haunted many scientific communities. How are terms different? What does each term cover? And how are all the gaps addressed by terms? These questions typically require long debate before consensus. Since I do not want to spend too much time debating with myself, I have endeavored to select pre-established terms whenever appropriate. If these terms and measures do not feel right for the systems being studied, please do debate to find better terms.

2.6.1   Quality Category 1: Ability to Sustain

Many of the sustainment metrics below come from industrial best practices for assessing systems. Though the scales may be different between physical machines, computer systems, and organic systems, the principles of quality remain similar. Human factors in system operations add a degree of complexity and variability to some of these metrics. Systems with human parts have to take into consideration the risk of unplanned behaviors, the evolving capabilities of people, and the display of secondary system characteristics with people self-organizing while performing functions. I will highlight some of the effects of human involvement in the quality measures below.

Measure: Operational Reliability. The percentage of time that a system will operate within intended parameters is a measure of system reliability. This can be computed by tracking the total time of unacceptable operations or nonoperations within a set period. Reliability standards can also be specified for activities conducted by the system, and the duration for each incidence of system failure can further be tracked. System operations can vary within allowable ranges for reliability. To enable systems to stay within this operational range, the parts can be designed to interact with minimal or no failures, redundancies in parts can be established, and backup parts can be prepared for rapid replacements. Sometimes, operations are adjusted when parts are not in use to minimize disruption and impact on output results.

Parts in a system might exhibit a pattern as they fail, which resembles a bowl-shaped curve. Along this curve, a higher number of parts fail early (infant death) due to manufacturing defects, and a higher number fail after a period of wear (senescence). Infant death in parts can be reduced with advances in manufacturing capability, and parts replacement can be planned after a certain period of use to delay senescence. When the failure characteristics of parts are not well known, the brute force approach for getting higher system reliability is to over-strengthen the parts for their intended functions. People, as parts in a system, add adaptability but present a challenge for determining reliability. Statistical factors in human performance can be affected by many subfactors such as the quality of new recruits, morale of the workforce, and reactions to external and internal events. Rigorous training and clear procedures can reduce variance. However, there are still no guarantees.

Measure: Accuracy in Performance. The accuracy of a system refers to its ability to achieve a specific set of performance results within very small variances. Like many other parameters, accuracy can be measured as a minimum acceptable threshold and a targeted objective. Accuracy can be achieved by design, or accuracy can be achieved by system adaptation and evolution. For example, the human mind and body are learning systems where the ability to perform tasks can be dramatically enhanced through practice, figuring out what has been done wrong, and making corrections in performance. Beyond mental learning, our muscles also recognize when each must get stronger to meet performance needs. Human organizational systems are also learning systems, as argued by many system thinkers. At the same time, managers of organizational systems still want to instill high accuracy by design. These ways to gain accuracy do not have to be mutually exclusive.

While the human mind and systems that mimic human learning can grow in performance, the advancements of computer control and high-precision mechanical devices have dramatically increased accuracy by design for physical systems. In contrast, accuracy in information systems generally refers to the fidelity of data, appropriateness of data, and lack of junk or erroneous data. Data accuracy in the age of high-capacity computing has launched a dedicated field of study into systematic and non-systematic errors with data as discussed earlier.

Measure: Affordability in Maintenance. If money is not a constraint, then any man-made system can, in theory, be maintained forever. In fact, at some point one will have replaced all the parts. The debate regarding when to stop maintaining the system is, thus, a consideration of affordability. Affordability is a trade-off among the financial, time, resource, and even emotional cost of maintenance versus the consequences of letting the system fail. In some cases, the cost and benefits of new systems with better technology vastly outweighs extending maintenance. In other cases, the risks of transitioning to a new system compel decision-makers to continue relying upon the old.

Systems can be designed for maintainability with features such as modular components, internal failure detection and warning, affordable parts, clear replacement processes, and methods for quick fixes. For organic systems such as the human body, maintenance is often a question of capability and not affordability in first-world nations. With the ability to replace organs, eradicate cancer cells, predict hereditary diseases, and alter the metabolism, the degree on which the human body can be maintained seems to depend primarily on the level of external trauma, genetic flaws, and the natural aging processes. Now even the natural aging process is being investigated and challenged by today's researchers. For organizational systems, maintenance is the cycle of recruiting, training, and assigning new employees as positions become available due to growth and departure of current employees. Some organizations also have processes that drive out less-performing employees to sustain quality and to promote continuous improvements.

Measure: Resilience against Threats. Most systems have some form of interaction with the environment that contributes to the failure of its parts. For the human body, there are carcinogens in the environment that promote cancer. For vehicles, the road wears out tires, and shocks, sun, and rain wear out the body.

For applications on the Internet, there are viruses, malware, and hackers that can corrupt codes, usurp data, and seize operational control. Beyond the continuous dangers of operations, the environment and other systems in the environment can pose specific and targeted threats at the system. A system's ability to maintain a continuity of operations when confronted by such threats is sometimes termed resilience. Thus, the resilience of the societal infrastructure has become a popular metric in disaster response planning [67].

Systems achieve resilience by being able to operate after being severely damaged, being able to prevent or avoid damages, and/or being able to make real-time repairs. To operate when damaged, a system can have redundant parts or have the ability to reorganize through the use of remaining working parts. To prevent damages, a system can have barriers against specific threats, parts that are hardened against specific forces and energies, and/or the ability to counter the mechanisms of approaching threats. For examples, missiles can be shot down, intruders can be captured, bombs can be detected, and computer hackers can be traced. A system's ability to detect threats also means that it might able to avoid the threat through evasion, relocation, or concealment. Finally, a system can make real-time repairs through the ability to adjust operations, switch parts, and reestablishing normal operating conditions. As the duration and nature of threats can be long and complex, systems that are resilient against threats may at times need a combination of the above techniques to maintain operations.

Measure: Capacity to Scale. Some systems have the luxury of operating in a completely steady-state environment. Most environments, however, impose fluctuating, escalating, or declining demands upon systems. For example, systems serving human groups will often face higher demands during peak hours, and systems addressing a popular need might suddenly experience escalating demand through market growth. To handle an oscillatory demand, the system must either have the continuous capacity to handle peak loads or have the ability to scale to adequate capacity during peak hours. To handle a growth in demand, the system must be able to either scale at the rate of growth or scale to the projected level of future demand in advance.

There are several strategies to scale capacity to match growth. The system can grow in size to yield higher volume of output. The system can increase the rate of operations to yield higher output per unit time. And the system can be replicated and coordinated to yield a higher combined volume of output. These scaling strategies have been used to grow information systems to the level of supporting millions of users in an enterprise, and these strategies have also been used in factory operations. Scaling often needs to be considered in conjunction with reliability, accuracy, and maintainable. This is because changes in system dynamics can introduce new stresses and even new vulnerabilities to threats.

While I have referred to capacity as system output, capacity can also be a metric of the communications and transportation to and from systems. With current computers being able to produce data at a rate faster than the rate the Internet can upload and download, network capacity described as bandwidth has been a defining factor in system performance. With the current productivity rate of regions in the global economy, shipping capacity has been a key factor in the world system. The scaling of capacity for designed systems is often a question of cost. How much invested cost to increase capacity can be recuperated through higher rate of sales and/or higher prices? The inability to recuperate such investments has lead to some markets, particularly those in poorer regions, to be stuck with antiquated systems with capacity unable to satisfy demands. The scaling of natural systems is connected with the adaptability of the system and the consequences if a system cannot scale or scale fast enough. By definition, if a natural system has survived, then its capacity is adequate. Adequate, however, may not be comfortable enough for some human conditions. Thus, some human societies across the world have not changed/scaled for thousands of years, and other human societies have gone from using spears to sending people to the moon.

Measure: Physical and Information Security. Security generally refers to the system's ability to detect, prevent, and stop intrusions. This is in contrast to the system's ability to defend against massive attacks. Systems that occupy physical space may require barriers, locks, sensors, and guards as components of physical security. As physical space is three dimensional, barriers need to address all possible directions of intrusion. Barriers provide limited security because for every material there is a tool that can cut through it. Nevertheless, if there is an entry point, then breaking down the door might be easier than breaking through walls. Modern locks have, thus, become very sophisticated in protecting doors. Access can be based on individual unique biological signatures such as retinas and fingerprints, no-replicable keys, and complex passwords. Beyond passive physical security, active means include sensors that detect particulate matter, cameras that detect across the electromagnetic spectrum, automatic weapon systems, traps, and guards who can respond to the actions of actual or potential intruders.

Security in cyberspace in many ways mimics security in the physical world. There are firewalls to block malicious intruders such as viruses, malware, and hackers. There are identity management and access control schemes to let in authorized users. Access control covers entering into an application, such as by password and physical token, and includes user privileges when they are in the application. One can scan for intruders in an application and on the network, and one can monitor for the effects of intrusion. Once detected, intruders can be isolated (quarantined), erased/deleted, or disabled. Intruders, such as hackers, whose physical presence is away from the system can still be traced back to the point of origin and be dealt with by physical means or reverse cyber infiltration. Absolute security is a theoretical impossibility as long as there are evolving and adapting intruders. Therefore, system security should either be considered a mechanism for limiting damages in a sustainment situation or a component of system competitiveness in defeating adversaries.

Measure: Ease of Repair and Restoration. When a system is failing to operate within required performance ranges due to damages, the time, cost, and resources required to repair the system and restore it to normal operations is another quality of sustainability. Time is divided into the period for repairs and the period for restoration. In systems such as a factory, the employees might move on to other jobs if broken machines require weeks to repair. Therefore, even when the mechanical parts of the factory have been fixed, time might still be required to hire and train new workers before the factory can return to normal operational capacity. The time of repair can be slowed by the complexity of labor that cannot be resolved by simply hiring more repair personnel, and the time of repair can be delayed by the finding and delivery of resources/parts. The cost of repair is then the labor cost plus the cost of resources. For every hour or minute that a system is not operating, there is also a cost in lost productivity. For systems and damages that cannot be repaired by man, there may yet be self-repair capabilities within the system. The human body, for example, can self-heal wounds, eradicate diseases through the immune system, and recover lost memory.

Some systems can be designed to facilitate rapid and affordable repairs. However, repairs might not be economical if system functions can be recovered through affordable replacement systems. In fact, the knowledge of when systems are better replaced instead of repaired could drive design decisions. In human organizational systems, repairing people as a part of the organization is sometimes done through counseling and corrective actions. Like physical systems, many companies have found that it is often easier to replace nonperforming employees. When employees have value due to the cost of their training, depth of their experience, and uniqueness of their skills, repair versus replacement becomes a trade-off decision. For example, the military typically has to invest millions to train a fighter pilot. Therefore, the first course of action when a fighter pilot is having performance issues is to figure out how to resolve the issue. In some cases, such as for celebrities, the person's experience and skills can be so rare that the entire system must adjust operations to compensate for the idiosyncrasies of the person just to get that person's benefits.

2.6.2 Quality Category 2: Ability to Compete

Competition is either the normal operating state of a system or infrequent events that the system must adapt to or overcome. Regardless of the frequency and intensity competition, the ability to compete can draw its quality measures from the principles of war as explained by Sun Tzu [68], Von Clausewitz [69], and other military strategists. The key feature of competition is that the system is in an environment with other systems. The environment impacts the nature of competition, and competition can further be governed by a mutual understanding of constraints/rules. A misunderstanding of constraints, such as one side's willingness to use a weapon that the other side will not, can lead to defeat but also disgrace for the victor. Also, differences between competing systems that result in asymmetries in engagement can expose vulnerabilities. We will explore the defeat of systems in conflicts to greater detail in Chap. 3 on how systems break. For now, let us see what strategists have taught us about measuring systems competition or conflict.

Measure: Ability to Outmaneuver. A system is largely defined by its interactions. When the interactions must compete with those of other systems and changes in the environment, then the question is how fast, how sharply, and how accurately the system can change its interactions to gain better competitive positions. This ability to maneuver embodies situational awareness, tight operational control, innate dynamic ranges, and sometimes the ability to adapt. In physical combat, a maneuver often involves the speed and radius of a turn, and a fighter jet that gets within the turn radius of the enemy can get an arch of fire. However, in group combat, a maneuver can also be used to change the formation and distribution of the group so that one can place a wedge into enemy forces, surround enemy forces, disperse enemy forces, as well as get behind the enemy forces.

In communications, maneuver is within the war of words where one debater can trap another in an indefensible position through logical arguments. In such a case, a maneuver becomes the speed of reasoning, range of ideas, and accuracy of words. Maneuvers in cyberspace can mimic that of physical space, but the terrain of cyberspace is incredibly complex, and the speed of movement is typically at the speed of light. Therefore, a successful offensive maneuver is about using the terrain to find places to hide, points of vulnerability, and access pathways. Defensive maneuvers in cyberspace, then, are about speed of intrusion identification, speed of code integrity validation, and range of security scans. For all types of systems, the simple rule is that if one cannot maneuver then one cannot win.

Measure: Ability to Capture Position. If a system is to oppose another system or the environment, then the position in which it engages greatly impacts the probability of success. The position can allow the system to attack the vulnerable areas of the adversary. The position can allow a higher rate and level of attack while hindering the adversary's counterattack. The position might simply be more defensible against adversaries. In land engagements, great historical emphasis has been placed on capturing the high ground. The high ground in traditional warfare allows the force to have greater visibility and greater range of fire. The adversary will typically face challenges in charging up the high ground, but the visibility of the high ground might also make it an attractive target for long-range weapons in modern warfare.

Find and getting into the right positions in conflict or competition is, therefore, ultimately about understanding advantages. So position is about the preferred state of the system in the environment relative to other systems. Systems stuck in disadvantageous positions can potentially change to the value of the position by shifting the nature of engagement. For example, a company selling a less desirable product in the market environment can potentially change the value of its position by rebranding the product instead of changing the product. Or, a military force trapped in a cave might find the cave to be a perfect hiding place if aerial bombardment can be brought down upon enemies surrounding the cave. The lessons-learned is that the advantages of positions can change rapidly in the course of competition or conflict. So, capturing positions is a continuous assessment and endeavor.

Measure: Ability to Concentrate Mass. In any engagement, the stronger side tends to win. In attacking specific points, the intensity of force instead of just pure strength tends to do more damage. Strength and ability to project strength are often connected with the concentration of forces to have the effect of mass. In military combat, concentrating mass means the ability of fighting units to come together and coordinate attacks as a group upon the vulnerable spots of the enemy. Long time ago when the accuracy of firearms was limited, concentrating mass literally meant bringing troops physically together. With today's high precision long-range weapons, concentrating mass is often about coordinating fire at a point. However, the idea of swarming for troop engagements in urban environments, as referenced earlier, is being modelled as a new way of combat. In nature, colony insects such as bees can suddenly swarm and attack to confront large adversaries. On social networks, participants can suddenly swarm around one person or event to overwhelm the situation. Awareness of the swarm can spread virally across the network to gather more participants to feed the intensity.

Self-formed masses raise the question of control. Centrally controlled masses can in some cases more easily maneuver and focus the energy of attack. Mutually coordinated masses might have more adaptability and agility against shifting situations. Finally, there is the idea of an uncontrolled and uncoordinated mass, which is essentially a mob. In a mob, each participant acts based on personal awareness, but the totality of the uncoordinated action can yield significant accumulated damage. Because of a lack of control and coordination, mobs can be hard to disburse unless all members of the mob are attacked. As noted earlier, mobs typically die down as the emotional intensity for creating the mob dies down.

Measure: Having Means of Offense. In systems competition, offense constitutes the set of actions and dynamic changes taken by the system to affect the operations, integrity, and performance results of one or more rival/enemy systems. The system can levy these actions directly upon the other systems in overt or covert attacks or upon the environment, which then translates into negative effects on the other systems. Offense brings together the results of situational awareness, capability to maneuver, and advantages of mass. The position then dictates when and whether offensive actions are likely to succeed.

Offense should start with the strategy that focuses on how to achieve the total end state outcome and extend down to the tactics and tasks. This traceability is specifically a part of US military planning as a formalize process. In general, strategies should be straightforward and may be inherent to the nature of the system. Then, the tactics and tasks can be complex with many interdependent adaptive steps as well as many options depending on the engagement situation. Some systems might have only a limited number of ways to conduct offensive actions, such as a single weapon for combat troops. In cases where the means of attack is obvious, tactics are particularly important in offensive success. A part of successful tactics is generally the element of surprise, which hinders the opponent's reaction time and ability to adapt. Other tactically elements include false information, fake maneuvers, withheld forces, timing of moves, and when to stop.

Measure: Having Means of Defense. In systems competition, defense constitutes the set of actions and dynamic changes taken by the system to protect its own operations, integrity, and performance results from one or more rival/enemy systems. Defense can be brought forward to the launching point of enemy attacks and can be established around each potential target of enemy attack. Defending the whole system can be more effective than defending all the system parts and associations, and layered defenses is often a preferred strategy to mitigate risks. When defending parts, one must often divide one's forces, which reduces the ability to overcome focused attacks.

Successful defense requires understanding the opponent's means of offense as well as an ability to predict opponent's strategies and tactics. Getting surprised by the enemy is a key failure of defense. Therefore, in engaging the opponent, defensive actions still need situational awareness, adequate mass, and the ability to adaptively maneuver. Passive defenses such as walls, traps, and automatic weapons can slow down attackers, but attackers must be actively engaged by counterattacking defenders in order for defense to succeed in system competition or conflict. While offense has the advantage of attack, defense has the advantage of home environment. Some environments, such as the Russian winter, are so harsh that they can be exploited by the defender to gain victory over massive forces such as Napoleon's army. Other environments favor defensive positions in engagements and maneuvers. However, defense alone will typically not yield total victory unless the enemies are willing to unrelentingly commit and lose their forces without retreat.

Measure: Gaining Situational Awareness. Both offense and defense require an understanding of the environment, the system's position and dynamics within the environment, and the opponents' position and dynamics within the environment. Therefore, the system's ability to gain situational awareness is a quality measure. The environment, in many cases, includes the physical domain of competition and the information domain of competition. Situational awareness of the physical domain will be in the form of sensory data, research data, data gathered from adversary sources, and interpreted information. However, the information domain of competition or conflict is about the system's dependence on information and the opponent's ability to attack that information to harm the system. If the system itself is composed completely of information and software that executes information, then attacks that erase data, corrupt data, steal data, disable operations on data, block the transmission of data, and generate false data can potentially destroy the system. Situational awareness information in this environment is both a measure of the system's competitive quality as well as a target of attack. Physical systems such as sensors can be attacked to hinder situational awareness. However, actual situational awareness exists in the information domain, and it is a part of the information system.

The value of situational awareness is based on how it enables the mechanisms of offense and defense. In fact, too much extraneous information about the situation can delay response time and decision cycles. This brings up the strategies for gaining situational awareness. One strategy is to sense as much as possible and filter out the data that give insight on how to attack and defend. Another strategy is to model the environment and opponents to find indicators that support competitive actions. Then, the sensor can focus on the indicators that matter in engagements. Finally, sensing can be based on predicting opponent behaviors and testing opponent reactions with probing actions. For example, if predictive models or observations seem to show that the opponents always turn right after an attack, then luring the opponent into a false turn can validate this understanding. These strategies can all be effective under different circumstances. What is most important in situational awareness is not just the right information but the right timing. If decisions must be made quickly, then the challenge of situational awareness is on how to interpret limited and/or imprecise information to allow for the highest probably of success. All those in the field of military intelligence understand the fog of war and the inherit risks in chaos.

Measure: Ability to Avoid Detection. The ability of a system to avoid detection in the physical and information space blocks the enemy's ability to accurately attack and may enable the system to gain the element of surprise in attacking the enemy. In physical space, avoidance can be by stealth technology, which renders the system undetectable to opponent sensors, or by evasive maneuvers, which places the system out of range or away from the angle of opponent sensors. For example, current stealth technology only makes crafts invisible to enemy radars and not to visual sightings. By the time the enemy can visually see the system, however, it will be too late to prevent long-range attacks. Similarly, many directional sensors, such as radars, have a speed of rotation for scanning 360°. A system that can maneuver within the scan cycle of sensors can launch surprise attacks.

Both in the physical and information domain, using the terrain to hide system activities can be an effective way for avoiding detection. Troops use camouflage in the wilderness. Viruses hide in animal cells. Computer worms mix into software codes. When a system has infiltrated the enemy's environment or even the enemy system itself, a means of avoiding detection is to mimic the behaviors and characteristics of enemy systems. For example, terrorists can hide in plan site and disguise their intent to succeed in attacks. When a system cannot completely avoid detection, hiding its more vulnerable parts can reduce the effects of being attacked. For example, organizational systems that are obviously competing in the market place will still have secret projects, false propaganda, and ways to hide activities within market chaos.

The diversity of means to avoid detection suggests that quality should simply be based on one's signature size relative to an opponent's sensors and the time as well as resources it will take for the opponent to figure out one's position and intent. Naturally being completely invisible during competition or conflict is an awesome capability. However, one only needs to be confusing enough to opponents/enemies in offense and defense to succeed.

Measure: Having an Economy of Motion. A system does not have to be efficient in using its dynamic capabilities to win competitions and conflicts, as long as it can overwhelm the opponent with brute force. This means that the system must have far more energy and perhaps expendable parts than the opponent. Unfortunately, a competing opponent that understands the inefficiencies of a much more powerful system can push that system to waste its strength and resources to gain victory. The competition for efficiency is fairly simple. In offensive and defensive activities, the measure is how much energy, time, space, and resources are needed by one side relative to another side in actions and counteractions.

Energy use as a measure of efficiency is straightforward because all physical and informational systems do not have unlimited sources of energy, and many systems operate under tight energy constraints. Fighter aircraft only have a few hours of fuel. A computer's consumption of electricity depends on level of processes. And humans can only work for so long without rest and nourishment. Therefore, efficiency is about meeting objectives with a pattern of dynamics that optimizes the use of energy. What is less obvious is the idea of time and space as measures of efficiency. Time is integral to system dynamics because all actions can be measured by rate and acceleration. The pattern of dynamics can increase or decrease the time a system takes to execute offensive and defense actions. Opponents might exploit delays caused by inefficiencies to gain competitive advantage.

The idea of space as a measure of efficiency is tied to the ranges of maneuver. Extremely large systems with a great deal of potential force might suffer from the inability to efficiently position themselves in competition. Therefore, there is, in many cases, a trade-off between levels of brute force and the economy of motion for accurately applying force. This trade-off is particularly important if resources other than energy are expended in competition. For example, a conflict between systems could result in the destruction of parts and the disabling of links. No matter how much attrition a system can withstand, there will always be a rate of damage that can kill the system. In the days of total confrontational warfare, military forces triumphed based on who has collapsed first from the loss of men and machines. Victory without mass destruction is perhaps the better way. Therefore, a system's ability to find the most efficient way to overcome the operations of an opponent could be the most valuable measure in competition and conflict.

2.6.3   Quality Category 3: Ability to Improve

The third category of quality measures pertains to a system's ability to change its capabilities, missions, and functionalities. In changing capabilities, these measures connect with the measures for sustainment and competition. In changing the functions of the system so that it can address new missions, these measures connect with the raw potential of the system stemming from its structure, boundaries, and interactions. Thus, these measures are either the summation of deeper understanding and predictive results about the system or a representation of goals and possibilities that the system wants to attain. In the latter case, the challenge is to find reasonable indicators for success without the need to wait for deeper systems understanding to manifest. Such indicators can be discovered through comparative systems analysis, historical trends, and connecting macro-system behaviors with the measures.

Measure: Flexibility in Dynamics. The range a system can adapt and evolve is expressed as flexibility. The system may not need to reach the maximum range of adaptive flexibility during normal operations and even during standard conflicts. However, this range helps us understand whether a system can expand its functions and take on more missions. In some cases, a system might be so flexible that it is wasted or poorly suited for its assigned mission. Then, the system's role in the environment should be reevaluated. In system evolution, flexibility can be used to describe the number and complexity of steps required to evolve from one structural and boundary state to another. Although unstable systems are more prone to change, it is really the range of change and the ability to reach objective states that determine flexibility. A system that can suddenly reorganize itself is simply unstable when the outcomes are chaotic and uncontrollable.

In order to increase flexibility during adaptation, a system could test the ranges of flexibility and try to figure out ways to extend the range. Just like stretching the human body, flexibility in other systems might also increase with hard workouts. In other words, the practice of adapting can increase adaptiveness. Another way to increase flexibility is to identify and modify parts and associations that are hindering flexibility. In extreme cases, a part can be shut down and a link can be broken to sacrifice system capabilities for system change. The most obvious sacrifice is to turn off the parts that are vulnerable to adversary infiltration because the threat is often more important than the contribution of the parts. When an objective state for flexibility is determined, the system can then be pushed to evolve to that state. This evolutionary step might be difficult when the system does not want to change. However, the breaking down of resistance will enable easier evolutionary shifts and contribute to achieving objective flexibility.

Measure: Agility in Dynamics. While flexibility is about the range of change, agility is about the speed of change. In the case of adaptiveness, speed applies to all four phases of the response cycle. Agility in observation is connected with the rate in which situational information is gathered. Agility in orientation is connected with the speed of assessing the situation. Agility in decision is connected with the speed of weighing all the options. Finally, agility in action is the speed in which the plan of action is translated into actual results. In the case of evolution, agility can be measured by the number of steps required to get to an objective system configuration and the time it takes to take each step. This time can be delayed due to system resistance and internal dynamic complexities.

To increase agility, the system can try to increase the flow of information and the speed of processing. However, reducing the required information, simplifying the assessments, and narrowing down the decision trees could have more dramatic effects. If such actions reduce the system's capabilities, then there is a trade-off between agility and capability. For example, is it better to fire more times from more angles in combat than to fire with higher accuracy? If the system cannot win the conflict with one strike, then maybe the agility of attacking with many strikes and the fog of war might yield better results. This type of thinking has entered into the software development process, where the idea of trying to get the perfect solution through initial planning is abandoned and replaced with the strategy of getting something working fast and then proceeding with rapid rounds of iterative development to reach a better solution. Agile software development, at the conceptual level, is, therefore, the argument for rapid system evolution as a better quality measure for improvement than all the measures associated with system design and operations. I will not take sides in this debate but will reemphasis that agility very often has trade-offs that depend on the situation.

Measure: Ability to Resolve Problems. If there is a problem with any of the quality measures discussed, then a system's ability to self-resolve the problem is a measure of the ability to improve. Traditional mechanical systems have limited self-repair capability and must rely upon redundancies and backups to deal with operational problems. In contrast, human organizations and systems with human components have problem-solving potential embedded at every layer of human involvement. Turning potential into capability is then the challenge of delegating authorities for making corrections, establishing accountability for empowered workers, enabling coordination to avoid chaos, and providing training to enable problem solving. The balance between centrally controlled correction and distributed real-time fixes is system and situation dependent. However, the fixes can be temporary to keep the system operational until more corrective actions can be taken, restorative so that the system is working as well as before the problem emerged, and preventive in that the system is changed in a way that similar problems are less likely to occur. For all these objectives, the problem-solving approach can be innovative. And innovation is sometimes driven by the necessity of the problem environment such as limited time to react, insufficient resources for repairs, and failure of standard processes.

The advancement of computer control systems has raised the question of how artificially intelligent can computers be at resolving system problems. If a computer can beat the best human chess player, can a computer cope with the unpredicted problems of system operations? By definition, real problems cannot all be predicted because known patterns of system failures can be handled as a part of the design process or system operation procedures. Computer artificial intelligence capability can handle newly emergent problems but might face challenges if the problem space is not well bounded like that of a chessboard. To rival the human mind in making sense of complex situations, computers need to advance beyond linear logic and into the ability to process the total situational information as a whole to match naturalistic cognition as discussed. Even so, that spark of innovation in humans is difficult to quantify and difficult to replicate.

Measure: Ability to Mitigate Risks. A system's understanding of what might happen in operations and competitions can be based on its ability to predict behaviors, learn from past experiences, and model/simulate outcomes of interactions. Predicting behaviors can yield an understanding of potential negative outcomes that have never occurred in the past. Learning from the past can yield an understanding of past negative outcomes that might reoccur in the future. Finally, the modeling and simulation of scenarios for when the system engages the environment can yield an understanding of ranges in potential futures. This understanding can be more than just the possibility of one bad outcome because the future is fluid. Instead, it can be of the possible paths in actions and consequences starting from the current state. Risk mitigation as a quality measure for improving the system is, therefore, not just about reducing probability of occurrence and level of negative impact. Mitigation can also be about shaping the future.

For most systems, the human element is what enables the mitigation of risks because the complexity of risks benefits from human judgment. The earlier discussions about part and association characteristics, system structure and boundaries, and system interactions help us to establish predictions and find ways to alter predictions. We can extrapolate trends while finding ways to stop the progression of trends and games through scenarios while finding ways to shift outcomes. In predicting the complex outcomes of system dynamics, agent-based models as explained earlier are powerful tools. In modeling human decision-making during the competition of systems, Game Theory from the economic and conflict science communities has yielded actionable results. The theory essentially assumes that each person and system will rationally act to preserve its self-interest. The decisions do not have to be perfect nor correct. However, they have to be rational because that is the basis of measure. The outcomes of back and forth rational engagements can reveal hidden patterns of risks. The potential of using complex system models to identify risks is that small adjustments in system operations might lead to vast improvements in risk reduction. Actions taken early will enable effects to proliferate, whereas responses to imminent risks may have to be far more aggressive. It is easier to stop a threat as it is forming, and it is easier to prevent a failure at the first instance of cracks.

Traditionally, analysts have portrayed risk in a reference of probability of occurrence versus the magnitude of impact with high probability and high impact being the risks that should receive the greatest mitigation focus. In the context of measuring a system's ability to self-mitigate risks, this framework can still be used to determine whether all the greatest risks have been properly discovered and mitigated. The term mitigation is used to discuss risks because risks cannot always be eliminated. Sometimes, the probability of occurrence can be brought down, and other times the level of impact can be reduced. If the ability to accurately model outcomes is not available, the mitigation of risks may have to focus on how to recover from the aftermath of negative events.

Measure: Ability to Capture Opportunities. The ability to identify ways a system can improve its structure, performance, and situation relative to the environment and other systems stems from the same capability as projecting risks. Thus, a system needs to understand the consequences of past behaviors, the possibilities of new outcomes, and the potential results of interactions. However, the capturing of opportunities deviates from risks mitigation in that the system needs to have an awareness of what it can become and what it wants to become. In risks mitigation, the system is still focused on its established mission/purpose and how operations, adaptations, and even evolved states can change to increase likelihood of achieving mission.

For a system to seize opportunities, it must have a vision that is beyond the current mission. There should almost be separation between the current mission and the vision because if the system can expand the mission to meet the vision, then the system is only an evolutionary or adaptive path. A system that is self-aware (senescent) can form this vision based on likes, hopes, dreams, and even fears. Then, an opportunity is what occurs in its operations, in the environment, and with other systems that makes a path from its current state to the vision possible. For example, a local store might want to be a global corporation but has no plan. Then, when it sees its products being spread to other cities by passionate buyers, the opportunity for a global brand suddenly emerges and can be captured if the system has the right qualities. Some systems are very good at scaling capacity but have low ability to seize what has not already been laid in as planned. In contrast, other systems can operate without a vision, see what is possible in the changing of situations, form the vision in moments of inspiration, and capture the opportunity to meet the vision. This quality is a corner stone of innovation, and the potential often cannot be predicted for systems based merely on existing capabilities.

The mystery of innovation is perhaps a good place to leave this section on system quality because I want to again emphasize how hard it is to fully understand nonrigid systems. If the system has human participants or is composed of human components, then the element of creativity and inspiration is always a part what the system might become. All statistical measures of human behaviors have outliers, and the outliers are in many cases the ones who change history. Therefore, systems that amplify human behaviors, such as armies, companies, political groups, and even mobs, can do surprisingly great and horrible things.

2.7   Integration: System of Systems

This last section of the chapter on how systems form is named "integration" for both the integration of the previous sections and the understanding of how systems form through integration. In the previous sections, I have taken a step-by-step approach toward exploring systems by starting from their parts and moving toward the system as a whole. Along the way, I have applied the concepts to examples of real-world system types. However, the question of how systems form still has not been clearly answered. This is because the answer extends back to the very origin of the universe and the origin of life. Even if one looks at a simple man-made mechanical system, there is the question of where the materials come from in nature. Then, if one thinks about the composition of each material, it is clear that each is formed from atoms, which are systems formed from elementary particles. The same can be said of human organizational systems because each human part is a complex system that is composed of complex cellular systems.

So as already suggested, all systems are at a level in the total system of the universe. We can then study a piece of the universe by picking a level of system description and a region of systems. Further, we can look below the level of study and above that level to understand how systems are composed of smaller systems as well as how systems belong to bigger systems. Our pattern of study in the prior

Fig. 2.35 Systems and larger systems

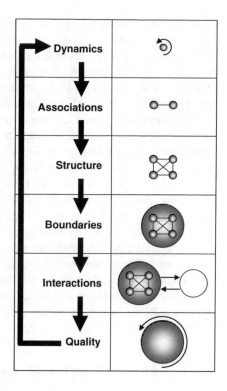

sections can, therefore, be iterated from understanding systems to understanding how those systems are mere parts to other greater systems, as shown in Fig. 2.35. At some point in this process, the details or the scope of study goes beyond our capabilities and perhaps beyond our needs. In fact, most system studies are narrowly focused, and it is only through recent advances in high-speed computational capabilities that researchers can start to model large complex systems.

Four groups of systems within systems can be defined to help us understand the levels we want to focus on in the activities of system formation. These groups are only to help us grasp a universe where systems are constantly being formed and systems are constantly breaking apart. The first group consists of systems formed by nature. The second group consists of systems formed by man. The third group consists of systems formed by nature using man. And the forth group consists of systems formed by man leveraging natural systems.

2.7.1   Natural System of Systems

These systems start with inorganic systems across the cosmos formed by energy and particles in space. The path of system interactions and transformations that has led to the creation of the life sustaining earth environment is still a mystery to scientists. Once the conditions for life exist, however, the second level of systems is the organisms in the environment ranging from virus DNA strands and single-celled prokaryotic life forms to complex eukaryotic multiple-celled organisms and, ultimately, the human system. I am not going to engage in the evolution debate. However, what is obvious is that the earth is teaming with life and that the great variety of life works together to form a biological ecosystem.

Living systems depend on one another for food, hosting, air content, internal biological processes, migration, and other functions. The human body, in particular, depends on bacteria to support digestion, plants for clean air, a great variety of plants and animals as food sources, and domesticated animals for work and transport in early civilizations. To be specific, many human societies would have turned out quite differently without the power of the ox and the capabilities of the horse. The next level of living systems working together is colonies of organisms such as bacteria, insects, and even mammals that form complex social systems. I have discussed some of these colonies earlier, but group behavior among animals can be more diverse and more complex than I have portrayed. When we see flocks of birds in the sky, packs of wolves in the forest, swarms of bees around a nest, and schools of dolphins in the sea, we can now think about how to model their system dynamics during normal operations and when they are threatened.

2.7.2   Man-Made System of Systems

These systems start with the basic mechanical machines that operate based on human, gravitational, electrical, chemical, and nuclear forms of energy and extend to electronic devices that control current telecommunications, computing, and sensing. Together, they form the infrastructure of society with systems all linked together by roadways, fiber optic networks, transmission towers, pipes, tracks, high-voltage cables, and transportation vehicles. Human society came about through centuries of learning, inventiveness, and dedication. Some parts of the infrastructure date back for decades and even centuries while others parts are constantly advancing. For example, some roads in the ancient city of Rome have been in use for the past two thousand years. In contrast, engineers are continuously trying to increase the communications capacity/bandwidth of wireless networks. When we take apart a computer device today, we literally see another world of designed systems within. This complex world is centered on the microprocessor, and if we peel open the microprocessor, we will see countless microscopic pathways and controls.

On top of the societal infrastructure sits systems of information. The earliest system of information is actually currency and currency-driven economic organizations. Currency is an information system because it tracks gained wealth, transfers wealth, and allows wealth to be reused for gains in the society. In the days before computing, paper currency and coinage enabled this information flow by physical transfer between hands and by communications between banks. Computer-based accounting and monetary exchanges then revolutionized system dynamics and now support the modern global economy. Other information systems include education and research, markets and sales, and coordination of social services.

2.7.3 Natural System of Systems with Man

These systems start with social groups formed by people from the earliest tribal societies to the modern social networks on the Internet. Although the Internet has allowed people to break past the barriers of distance and divided communities, the interactions between people have remained largely the same for thousands of years. At the end of the day, people connect to seek friendship, find love, rally against common enemies or concerns, discover common causes, and exploit others. If there is a common cause, which could be as simple as surviving as a group, then the system can become more structured with leadership, roles and responsibilities, and rules and procedures.

While people may see that the way in which we can organize ourselves is based on human rationality and ingenuity, we often forget that our minds are natural products. Therefore, we cannot say that we fully understand human systems because we still do not completely understand the mysteries of the mind (thoughts, dreams, and emotions) and the functions of the brain that sustain the mind.

The system of systems that is perhaps the most challenging to understand are those human networks that have spread across vast regions with millions of participants. Millions of people cannot all know one another nor can a system structure assign individual responsibilities to millions of people. Though the purposes of people interacting may still be simple, there could be layers and layers of self-formed groups and subgroups. One person can belong to a variety of groups that have system behaviors, and thousands of systems can be forming and reorganizing within the overall system of the network. As one might start to wonder whether such vast networks are really whole systems, massive adaptive group behaviors across the world can emerge to surprise researchers.

2.7.4 Man-Made System of Systems Leveraging Nature

Finally, there are systems created by man through the manipulation of natural systems. This manipulation refers to more than just man using materials from nature in simple and sophisticated ways. The first type of natural systems that man can manipulate is the inorganic systems of nature. For example, the atom is a natural system that enables the formation of molecules and materials. Man has long learned how to create chemical reactions that alter the arrangement of molecules, and man has recently learned how to trigger certain types of atomic reactions. However, these are still blunt-force approaches to merely harnessing energy from molecular and atomic level bonds/links. As we start to learn how to control molecular structures to form nano-systems, and, as we start to figure out how to conduct controlled atomic alterations, science will have progressed to truly leveraging nature systems.

Ironically, our ability to leverage organic systems is perhaps more advanced because of the size and coded structure of the DNA molecules. As DNA is the blueprint for all known life, scientists have discovered how to map DNA and how to alter DNA to change the behavior of cells and the production of proteins. Having a map and knowing how to change the map to alter system dynamics is not the same. Thus, much advancement is still required for scientists to fully manipulate organic systems. For example, we can now genetically engineer bacteria to yield useful byproducts, and we can genetically engineer plants to be more disease resistant and have higher yields. Yet the genetic manipulation of complex animal systems is fraught with dangers as well as ethical issues. First, we do not fully understand the cellular reference frame in which DNA codes are applied. Then, the functionalities of many gene sequences are still unknown to us, even though the codes of all the DNA strands have been mapped. If we change a known gene sequence, there could be hidden consequences and proliferated effects as proteins interact with other proteins. To truly design new organic systems, all the ways that DNA sequences can be expressed to enable cellular and intercellular functions must be unraveled. The most challenging organic system or component to design is perhaps the brain. If man can start to design thinking systems, then what are the limits of systems evolution? Will the organic creations of man start to enhance themselves and reason about their existence? Such possibilities are within the scope of systems exploration.

As I close this chapter on how systems form, our exploration of systems and systems formation will continue through the specific study of system failures. The dynamics of how a system will break down can give more insight on how the system was built. For the process of breaking can be treated as a decomposition or destructive test of a system. Sometimes natural systems are so complex or so resistant against probes that their mysteries will not reveal themselves until breakdown starts to occur. Other times, the flaws in man-made systems might not be discovered until the systems are tested to the point of failure. What I have presented so far is only a foundation for systems thinking and systems analysis. Our journey of discovery continues.

References

1. Ren C (2000) Additional factors governing specialized cell dynamics based on deductive systems analysis. Theor Biosci 119(2):95–103
2. Rokach L, Maimon O (2008) Data mining with decision trees: theory and applications. World Scientific Publishing Co Inc., Singapore
3. Alpaydin E (2010) Introduction to machine learning. MIT Press, Cambridge

4. Chen Z (2001) Data mining and uncertain reasoning. Wiley, New York
5. Draper NR, Smith H (1998) Applied regression analysis, 3rd edn. Wiley, New York
6. Piatetsky-Shapiro G (1991) Discovery, analysis, and presentation of strong rules. Knowledge discovery in databases. MIT Press, Cambridge
7. Bailey K (1994) Numerical taxonomy and cluster analysis. Typologies and taxonomies. Sage Publication, Thousand Oaks, p 34
8. Kantardzic M (2003) Data mining: concepts, models, methods, and algorithms. Wiley, New York
9. Knorr EM, Ng RT, Tucakov V (2000) Distance-based outliers: algorithms and applications. VLDB J Int J Very Large Data Bases 8(3–4):237
10. Bojadziez G, Bojadziez F (1997) Fuzzy logic for business, finance, and management. World Scientific Pub Co Inc., Singapore
11. Castillo E et al (1997) Learning Bayesian networks. Expert systems and probabilistic network models, pp 481–528
12. Pawlak Z (1982) Rough sets. Int J Parallel Prog 11(5):341–356
13. Banzhaf W et al (1998) Genetic programming—an introduction. Morgan Kaufmann, San Francisco
14. Bishop CM (1995) Neural networks for pattern recognition. Oxford University Press, Oxford
15. Gilbert N (2008) Agent-based models. SAGE Publications, Thousand Oaks
16. Wickens CD, Hollands JG (2000) Engineering psychology and human performance, 3rd edn. Prentice-Hall, Upper Saddle River
17. Saaty TL (2001) Fundamentals of decision making and priority theory. RWS Publications, Pittsburgh
18. Thaler DE (1993) Strategies to tasks: a framework for linking means and ends. RAND Corp, Santa Monica
19. Storkerson P (2010) Naturalistic cognition: a research paradigm for human-centered design. J Res Pract 6(2):12
20. Van Wylen GJ, Sonntag RE (1985) Fundamentals of classical thermodynamics, 3rd edn. Wiley, New York
21. Bunge M (1963) A general black box theory. Philos Sci 30(4):346–358
22. Stephen I (2003) The forgotten genius: the biography of Robert Hooke 1635–1703. MacAdam, San Francisco
23. Holtzapfel W (2014) The human organs: their functional and psychological significance: liver, lung, kidney, heart (3rd revised ed.). Floris Books, Edinburgh
24. Breay C (2010) Magna Carta: manuscripts and myths. The British Library, London
25. Clark DK (1954) Casualties as a measure of the loss of combat effectiveness of an infantry battalion. Technical Memorandum, Johns Hopkins University, Operations Research Office
26. Riché P (1978) Education and culture in the Barbarian West: from the sixth through the eighth century. University of South Carolina Press, Columbia
27. Easton TA (2008) The 3D trainwreck: how 3D printing will shake up manufacturing. Analog 128(11):50–63
28. Lécuyer C, Brock DC (2010) Makers of the microchip: a documentary history of Fairchild Semiconductor. The MIT Press, Cambridge
29. Drexler E (2013) Radical abundance: how a revolution in nanotechnology will change civilization. PublicAffairs, New York
30. Paprotny I, Bergbreiter S (eds) (2014) Small-scale robotics from nano-to-millimeter-sized robotic systems and applications: first international workshop. Springer, Karlsruhe, Germany June 16
31. Edwards SJA (2000) Swarming on the battlefield: past, present, and future. RAND Monograph MR-1100. RAND Corporation, Santa Monica, CA
32. Maslin M (2014) Climate change: a very short introduction, 3rd edn. Oxford University Press, Oxford

33. Miller MB, Bassler BL (2001) Quorum sensing in bacteria. Annu Rev Microbiol 55:165–199
34. Nelson DL, Cox MM (2005) Lehninger's principles of biochemistry, 4th edn. W.H. Freeman and Company, New York
35. Pohlman MB (2008) Oracle identity management: governance, risk and compliance architecture. Auerbach Publications, Boca Raton
36. Smith GL et al (2013) Vaccinia virus immune evasion: mechanisms, virulence and immunogenicity. J Gen Virol 94:2367–2392
37. Saini R, Saini S, Sharma S (2010) Nanotechnology: the future medicine. J Cutan Aesthetic Surg 3(1):32–33
38. Gilpin R (1981) War and change in world politics. Cambridge University Press, Cambridge
39. Mearsheimer JJ (1995) A realist reply. Int Secur 20:82–93
40. Waltz KN (1990) Theory of international politics. McGraw-Hill Companies, New York
41. Baldwin DA (ed) (1993) Neorealism and neoliberalism: the contemporary debate. Columbia University Press, New York
42. Whelman A (1994) Wilsonian self-determination and the Versailles settlement. Int Comp Law Q 43(1) June
43. Ratzel F (1897) Studies in political areas II: intellectual, political, and economic effects of large areas. Am J Sociol 3(4):449–463
44. Wendt A (1992) Anarchy is what states make of it. Int Org 46(2):391–425
45. Russet B (1993) Grasping the democratic peace: principles for a post Cold-War world. Princeton University Press, Princeton
46. Bull H (1977) The anarchy society: a study of order in world politics, 2nd edn. Columbia University Press, New York
47. Mackinder HJ (1904) The geographical pivot of history. Geogr J 23:421–437
48. Spykman NJ (1938) Geography and foreign policy I. Am Polit Sci Rev 32(1) February
49. Wallerstein IM (2001) The end of the world as we know it: social science for the 21st century. University of Minnesota Press, Minneapolis
50. Rostow WW (1959) The stages of economic growth. Econ Hist Rev 12(1):1–16
51. Larrain J (1989) Theories of development: capitalism, colonialism, and dependency. Polity Press, Cambridge
52. Smith A (1991) [1776] The wealth of nations. Prometheus Books, Amherst
53. Malthus TR (1990) [1798] Essay on the principle of population. Cambridge University Press, Cambridge
54. Ricardo D (1996) [1817] Principles of political economy and taxation. Prometheus Books— Great Minds Series, Amherst
55. Marshall A (1949) [1890] Principles of economics, 8th edn. Porcupine Press, London
56. Rosentein-Rodan P (1943) Problems of industrialization of eastern and Southern Europe. Econ J 53:202–211
57. Von Hayek F (1945) The use of knowledge in society. Am Econ Rev 35:519–530
58. Schumpeter JA (1984) [1942] Capitalism, socialism, and democracy. HarperCollins (Harper Torchbooks), New York
59. Reijinders J (ed) (1998) Economics and evolution. Edward Elgar Publishing, Cheltenham
60. Keynes JM (1989) [1936] The general theory of employment, interest, and money. Harcourt Brace Publishing, Stamford
61. Friedman M, Schwartz A (1963) A monetary history of the United States 1867–1960. Princeton University Press, Princeton
62. Marx K (1992) [1867] Capital: a critique of political economy vol 1. The Penguin Group, New York
63. Khimm S (2011) Who are the 1 percent? The Washington Post, Oct 06
64. Bar-yam Y (2003) Dynamics of complex systems. Westview Press, New York
65. Coram R (2004) Boyd: the fighter pilot who changed the art of war. Back Bay Books, New York

66. Hastings A (1996) Population biology: concepts and models. Springer, New York
67. Woods D (2006) Essential characteristics of resilience. Engineering: concepts and precepts. Ashgate, Aldershot
68. Sun Tzu (reprint 2010) The art of war. Tribeca Books, New York
69. Von Clausewitz C (reprint 1984) On war. In: Howard M, Paret P (eds) Princeton University Press, Princeton

Chapter 3
The Characteristics of Systems Breakdown

Abstract This chapter leverages the conceptual framework for how systems form to explain the many ways systems break down. These ways are organized into major methods, from violent destruction to intentional retirement, and specific breakdown mechanisms within each method. To explain the mechanisms, causes are conceptually presented, patterns of failure are diagrammed, and specific examples from society and history are introduced. Though the modeling of system structures can show the detailed behaviors of breakdown, the focus of this chapter is to help readers initially define the nature of breakdown concerns for specific systems so that resource intensive modeling and simulation efforts can be focused and productive.

In the real world, we often see the breakdown of systems more dramatically than the formation of systems. Systems formation can be subtle and slow, requiring generations of subsystem and configuration refinements before reaching a stable desired state. On the other hand, the same system that took generations to create and/or years to grow in the environment can breakdown in the blink of an eye. Think of the complexity of a modern jet aircraft and what it took to design and manufacture. Then, think about what is left after a crash. More importantly, think about the preciousness of human life on a commercial jet, each with self-identity, a lifetime of unique experiences, a circle of loved ones, and untold potentials. These lives, with genetic origins extending back across countless generations, can be ended tragically with a midair disaster. The philosophical reason for this dichotomy is that systems formation, no matter how complex, has to contain some degree of order. In contrast, systems breakdown can be orderly or it can be chaotic. The system that took steps to build can be smashed apart by random blows. The degree of order and chaos in system breakdown is connected with the method and purpose of each breakdown.

Not all breakdowns are sudden and dramatic, and even dramatic breakdowns might have hidden patterns of fractures and weaknesses extending back in time. Therefore, I will propose some common breakdown methods for the purpose of helping to frame the sections and explorations in this chapter. These are:

© Springer International Publishing Switzerland 2017
C.H. Ren, *How Systems Form and How Systems Break*, Studies in Systems, Decision and Control 72, DOI 10.1007/978-3-319-44030-9_3

- System broken by Conflict: Intentional damages to parts, associations, structure, and/or processes.
- System broken by Growth: Unintentional damages to parts, associations, structure, and/or processes.
- System broken by Decay: Unnoticed damages to parts, associations, structure, and/or processes.
- System broken by Obsolescence: Emerging ineffectiveness of parts, associations, structure, and/or processes.
- System broken by Stress: Forces that break down parts and associations during operations.
- System broken by Assimilation: Lost of structural separation, behavioral independence, unique functionalities, and system identity.
- System broken by Flaws: Errors in design and weaknesses in formation.

The above ways for systems to breakdown is a fair approach for capturing breakdown characteristics, studying the nature of breakdowns, and organizing the following sections.

I intentionally used the word "breakdown" because breakdown is very different than system failure. Sometimes, a system is broken down so that it can be reorganized into a better system. Other times, the breakdown is to make room for other better systems. Researchers sometimes intentionally break down a system to learn about its composition. And systems are broken down because they are doing more harm than good. None of these situations automatically mean that a system or class of systems has failed. Failure is a metric that links systems formation to systems breakdown. A failure has occurred if a defined mission for the system during formation cannot be met or an intended lifespan for a system cannot be achieved. Failure often implies a design flaw in the system, a construction or growth defect, or an inability of the system to function in an environment. Systems do fail, but not all broken systems are failures.

Because data on systems breakdown are sometimes readily available or are obtainable through experimentation, the study of real-world systems can start with how systems break and then proceed to how systems form. The reason that I have elected to start with discussing systems formation in Chap. 2 is to avoid being trapped by the limitations and ranges of data. There are dimensions of formation and even dimensions of breakdown that are not easily measurable. Nevertheless, we cannot ignore such behaviors in systems research. As I have tried to cover the full scope of characteristics in systems formation, I will endeavor to present systems breakdown, not based on current data and research results, but on the full scope of how a breakdown can happen.

Toward the objective of seeking out all the ways and patterns of systems breakdown, the framework of characteristics established in Chap. 2 of the book, such as for structures, boundaries, and interactions, will be helpful in defining the characteristics of breakdown. This will become evident as we simply jump into the methods and reasons for breakdown.

3.1 Conflict: The Intent to Damage

In exploring systems interactions as a part of formation, we looked at competition and conflict as a capability of the system in engaging other systems in the environment. Conflict in the context of system survival means the ability to defeat other systems to gain resources, territorial advantage, and operational advantage. And adaptiveness is a key characteristic that enables success in conflicts. What we did not look at is the ugly side of the consequences resulting from defeat. While defeat does not always mean system breakdown, there are only two ways to end conflicts. First way, one side or both sides have to yield through surrender or negotiated peace. Second way, one side or both sides have to be brought to a state where they can no longer engage. In the second way, the containment of a system will end the conflict, but the disabling of a system is often the approach of choice by all sides. This is the point where the concept of system conflicts converges with the study of how systems break. For the mission of disabling opposing systems is closely connected with actions to intentionally inflict damages on parts, associations, structure, and/or processes.

We live in a violent world, and systems in nature as well as those made by man conflict all the time. Our bodies are periodically attacked by bacterial and viral systems. Our lifestyles are threatened by criminal actions. And our society will at some point engage in wars with other societies. Even with the limitation of laws, adversaries in the business world might try to destroy us. Even with the bond of family and friendship, conflicting interests might turn us into enemies. Thus, the study of conflicts is perhaps the most popular path in studying the breakdown of systems. In the social sciences, the study of conflicts has focused on causes. In the military sciences, the study of conflicts has focused on strategies and tactics. We have reviewed the principles of war in assessing a system's ability to compete. In the causes of wars across the world and conflicts within societies, theories that could provide explanations include:

- Marxist Theory: Conflicts caused by wealth inequalities among people in society [1]
- World Systems Theory: Conflicts caused by regional inequalities in production and trade [2]
- Race Conflict Theory: Conflicts caused by inequalities and animosities between people of different races [3]
- Post-Structuralism Theory: Conflicts caused by the instabilities of human society due to the complexity of human agendas [4]
- Post-Modernism Theory: Conflicts caused by the imperfections of science and technology in achieving peaceful modern societies [5].

Beyond theories, we know that wars and social chaos have emerged due to religious differences, cultural differences, ideological differences, and simple political offenses. In fact, the causes of conflicts can be generalized into the

Fig. 3.1 Notional
representation of general
causes for conflicts

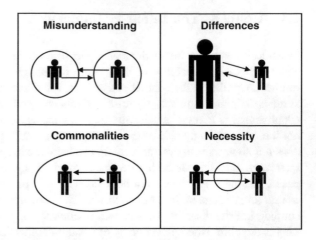

categories of misunderstanding, differences, commonalities, and necessity as
expressed in Fig. 3.1.

Conflicts caused by misunderstanding can be traced to actions based on mistrust,
reactions due to miscommunications, and decisions based on wrong detection or
awareness of adversary intent. These are obvious reasons, but we often forget how
easy it is to slip into misunderstanding. Misunderstandings are perhaps more likely
to occur in systems controlled by a singular dominant human leader. When the
decision of one person can launch a country into war, then the mistrust of that
person and how that person interprets communications become a critical matter. In
contrast, automated systems making decisions based on human established rules
and artificial intelligence engines based on more complex decision algorithms might
also make mistakes on when to attack. Even when the rules and algorithms are
perfect, mistakes can be caused by errors in detection and assessments of situations
involving other systems. Such is the concern of deploying automated weapon
systems and robotic guards.

Mistakes in automated controls suggest that we can perhaps place all accidents
in the category of misunderstanding. Systems can come into conflict with one
another simply because of a lack of awareness and not just because of decisions
based on mistrust. Planes can crash into one another because of faulty radars.
Armies can run into one another in the night and start firing before deciding to fight.
And one system, such as a nuclear reactor, can fail in a way that harms many other
systems. Though unintended, the outcomes of accidental conflicts can be as violent
and destructive as intentional conflicts. I should note that the studies of human
decision-making, artificial intelligence, and automated controls are very different
avenues of research. Even within these avenues there are branches and
sub-branches of study. Nuances such as deception between opponents, rational and
irrational thought, cultural barriers, and mathematically based control theories can
takes volumes to explain.

Conflicts caused by differences can be traced to religion, culture, and ideology. Further, wealth, power, and race are other forms of differences as noted by theories. While theories are looking for data to justify the causes of specific events, system behaviors suggest that any characteristic of a system that differentiates it from other systems can be a cause for conflict. This is because systems that cannot integrate but have to exist in a common environment will more likely compete. If religious, cultural, ideological, economic, political, and race differences do not create differences in the structure and operations of systems, then we might consider that such differences are less likely to cause conflicts. However, it is difficult for religion and culture to avoid shaping the fabric of the social system. Laws, societal investments, education, and the attitudes of the workforce are all quite vulnerable to the influences of religion and culture. Wealth and power also inherently change the social systems, as they create divisions for parts within a system and for groups of systems. The rich and poor in a society do not have the same functions and responsibilities. The more prosperous countries do not have the same social structures as the poor countries. The poor and powerless might be motivated to overthrow the rich and powerful in conflict. The rich and powerful might be motivated to suppress or even enslave the poor and powerless in conflict.

Conflicts caused by commonalities are driven by systems understanding one another perhaps too well. Systems with common characteristics have the potential to integrate. However, until that integration has occurred, each will understand the strengths and harmful potentials of the other. This fear can lead to preemptive strikes, and mutual understanding of vulnerabilities can lead to great damages with each blow. Even when fear is controlled, common structures and processes often yield the desire to compete for common resources and common territorial claims. Desire is the right word because the systems might automatically go after the same things even when there are enough distributed resources and territory to be shared. For automated systems operating under set rules, they will conflict when in the same environment unless either the capability to establish compromises or pick the second best choice has been built in. Conflicting systems will trigger and feed one another's common capacity for aggression until battles start.

The final conceptual cause for conflicts is simply necessity. Systems do not necessarily have to fight because of misunderstanding, differences, or commonalities. However, systems might have to fight to survive. When a bacteria or virus invades the system of the human body, the body must defeat the invader. When a predator sees prey, it must destroy and consume the prey for nourishment. Since the start of human civilization, man has dismantled natural systems to build systems of his own. In our conflict against nature, some have argued for balance to preserve enough nature for the next generation, and others have argued for restraint out of respect for nature. However, few have said that man cannot disturb the natural systems at all. In the global human society, man does not have to conflict with man. However, when one is forced to defend oneself from hostilities initiated by others, then one's decision to fight back is out of necessity. Sometimes we must act first out of necessity. For example, if we know that the first attack of an enemy will destroy us and we know that the enemy will attack, what should we do? The entire emphasis of the Cold War from the direct line between the White House and Kremlin to the early warning radar systems is to ensure that the enemy's intentions are not misunderstood [6]. In another example of striking first, if we know that deadly drug experiments on animals can save the lives of human beings, what should we do? In labs across the world every day, researchers are destroying organisms to learn about their structure and experimenting on mice, monkeys, and other animals to give humanity a fighting chance against diseases, toxins, and other threats. Even to ensure the quality of machines, a select number of products rolling off the assembly line must be destructively tested. Cars are crashed into walls with test dummies to measure their safety, and electronic devices are thrown to the ground to see how intensely they break. These actions may not seem like conflict, but they are still examples of man destroying out of necessity.

There is much contention in the academic world about predicting when conflicts will start and how conflicts will end. In my general categorization of conflict causes, I, by no means, want to get caught on any side of a specific debate. Therefore, I am intentionally staying away from specific examples, as almost any example I pick will have arguments and counterarguments. Even if I were to stick with the general categories for conflicts, some might argue that the cause is misunderstanding, while another might argue that the cause is necessity. Across the world today, history is written in many different ways by the victors and those defeated.

Setting aside causes and how systems interact in conflicts, conflicts can break down systems in several general ways. These ways of breaking down systems through conflict are represented by notional configurations, as shown in Fig. 3.2. They are then explored using our understanding of systems dynamics as revealed in earlier sections. Some ways of breakdown are linked to specific types of conflicts. If so, I will try to tie the mechanism of systems breakdown to the nature of causes and conflict strategies. Some ways of failure, which is an appropriate term for breakdown as a result of conflict, can be linked to a variety of conflicts. In such cases, I will try to explore the nuances of the mechanism that makes it flexible and broadly effective.

3.1.1 Breakdown by Crippling Strike

In highly integrated systems with strong dependencies on specific parts, the destruction or even the temporary disruption of key part or parts can break down the system. In the human body, for example, stopping the blood flow or the supply of oxygen to the blood will immediately cut the supply of oxygen to the cells. The brain will shut down and brain cells will start to die in three to four minutes. The muscles will lock and die in hours. And the body will cool down within 24 h. Therefore, ways to strike precisely to kill the body includes stopping the heart, cutting a major artery to drain the blood, suffocating the lungs, and damaging the brain enough to stop the heart. Of these attacks, the piercing of the heart through actions, such as an assassin's bullet, is perhaps the most immediate way to kill the

Fig. 3.2 Ways conflicts can break down a system

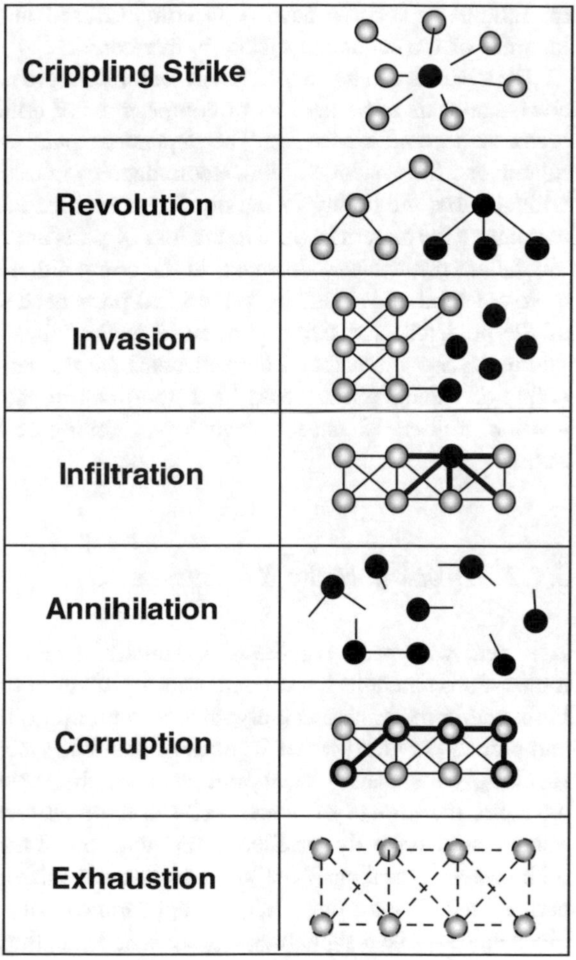

Crippling Strike	
Revolution	
Invasion	
Infiltration	
Annihilation	
Corruption	
Exhaustion	

body as people can hold their breath for minutes, try to plug or cauterize a cut artery, and live in a brain-dead state through machines.

In the human societal system, stopping the flow of electrical power and destroying the water supply can severely hinder processes. Thus, critical infrastructure locations, such as power plants, power relay stations, and water treatment plants, are targets in a modern warfare strategy that seeks to cripple the social system. The economic system, in contrast, depends on reliable and secure storage plus transfer of information. To avoid single point failures, data centers have backup options, processes are distributed, and technologies are redundant. However, the one unavoidable weakness of the system is the direct link between economic entities and electronic records. This permits criminals and national adversaries to attack the records through theft and data corruption. Whenever tens of millions of records have been compromised in a precision cyber attack, the integrity of the economic system is threatened.

The systems can be very different, but the failure mechanism of a crippling strike always appears to be the loss or compromise of critical parts. Then, the links will break or become ineffective. The dependent parts will become operationally disrupted, and the system will fail. Redundancies in critical parts, higher protection of critical parts, the ability to rapidly fix or replace broken parts, and the ability to restructure the system to survive the loss of parts are among the strategies to oppose this failure mechanism. However, in the competition between systems, there seems to be a natural understanding that critical parts need to be attacked. This is the way of the predators and the way of warfare. For those studying this mechanism of failure, tracing the paths of dependencies across parts and associations to singular points of vulnerabilities could be a good starting strategy. Alternatively, one can test the proliferated impact of each part failing or a small combination of parts failing.

3.1.2   Breakdown by Revolution

A system with parts capable of independent reasoning, able to operate autonomously, or vulnerable to external control/influence can face an attack from within. Parts within the system can organize into an opposition force against the structure and processes of the system. The result could be system breakdown, even when the revolting parts merely want to transform the system. In the human body, for example, the growth of cancer cells is a revolt within the system. Cancer cells operate autonomously, proliferate rapidly, spread across the system, and interfere with system operations. The ways to combat this revolt is to identify the cancer earlier based on the unique appearance of the cancer cell and the biochemical/protein signatures it produces. Then, the attack on cancer cells is either by precision, such as surgery and radiation beams, or identifying unique vulnerabilities, such as the propensity to absorb select chemical toxins or radioactive materials more than surrounding cells.

In human social systems, revolutions can emerge to oppose the established government or controlling processes. Intelligent rebels hide from authorities while recruiting and organizing, and their stealth as well as their understanding the system structure provides advantages in surprise attack. However, the process of evading detection also makes it difficult for rebel forces to buildup great capability. Initially fighting in an environment controlled by the system also carries clear risks. Thus, the rebel strategy is often to first break apart the system and claim one portion as its system. Then, the struggle becomes a civil war over control of the whole system. Rebellions can receive external support or can even be instigated by external advisors/agents. In such cases, the link between rebels in the system and supporters outside the system must be strengthened at the onset of revolution. The system, on the other hand, must seek to strangle this connection to block the flow of supplies, forces, and knowledge. Beyond attacking rebel parts one by one, the system can oppose a rebellion by attacking the specific types of association that are enabling rebel structures. For example, if an ideology is discredited or a means of communications is blocked, will the rebel organization disband? In complex rebel structures within the systems, there may also be critical points vulnerable to precision crippling attacks. Though the rebel leaders are obvious choices, other hidden vulnerabilities might be the one person who knows how to sustain cash flow or the one person who knows how to build weapons.

Revolutions are typically extremely violent because the conflict starts within the boundaries of the system and the opposing sides all have some understanding of the system. In most successful revolutions, there seems to be a point where the momentum shifts in favor of the rebels. Prior to the point, the rebels are continuously under threat of complete annihilation. The damage resulting from defeating rebels can be so great or the remaining system inherited by the rebels can be so torn apart that the system will quickly fail due to external stresses. For those studying this mechanism of failure, identifying the degree in which rebels can organize without detection and the way in which people will decide to join the rebels is a starting point. Alternatively, one can test the processes of the system against attacks from within the system, which uses means acquired in the system.

3.1.3 Breakdown by Invasion

The most straightforward conflict between two or more systems is that of system boundaries contesting one another. The invading system is the side that extends its boundary through force to break through the boundary of the opposing system. Once the opposing boundary has been breached, the invading forces can attack parts and links in the system to cause failure, demand surrender, and/or assume power. The invading forces in a system can still face strong internal resistance such as that faced by disease after breaching the human skin. However, the invader will have distinct advantages unless great losses can be inflicted on invading forces or great harm can be levied back across the invading boundary, perhaps along the path of retreating forces.

The most common forms of invasions in this world are with armies of men, armies of insects, and armies of plant life. In current and past human warfare, weapons and machines are used to transport men into battle and help men breach defenses along the boundaries of opponent systems. However, it is the soldiers and commanders that interact with the opposing system and decide which parts to attack and which links to cut. Computers can help commanders make decisions, but we have yet to find a replacement for the human component in adapting to the chaos of the battlefield and the ingenuity of adversary forces.

Animal invaders do not have the intelligence of human armies. However, many have developed advanced abilities through sound and movement to coordinate across vast numbers. This then allows the total army to achieve fairly complex results, even though the range of behaviors for one soldier can be easily modelled. Insect species such as ants are very representative of natural invading systems, and some ant populations can completely ravage a region. A region can also be completely dominated by invading plant life. Seeds not indigenous to a region can grow out of control with a new climate and without other plants that can successfully compete with the invader for soil and water. While plants cannot move and react like animals, their ability to leverage wind, water, and animals to carry vast quantities of seeds leads to a formidable force.

Those wishing to study this mechanism of failure can probably start with an agent-based model that simply represents invading units and defending system boundaries, parts, and links. This model will not predict exact outcomes, but the interactions of the agents can lead to the discovery of vulnerabilities. After multiple iterations and seeing how the system has failed, consistent patterns might start to emerge. Invaders with highly intelligent behaviors can be modeled through the various approaches in cognition theories. This could result in more complex engagements between invading units and system components, as broader ranges of strategic and tactical options can be explored. Classic failure patterns during invasions include the inability of distributed defenses to block concentrated attacks at one point, steady wearing down of boundaries with waves of attacks, exploitation of weak points in the boundaries by attackers, finding ways to cross the boundaries, and outmaneuvering system defenses at the boundaries.

3.1.4 Breakdown by Infiltration

When an invader is able to enter a system without attacking, then such an invader has typically managed to avoid detection by the system. The ways of infiltration include:

- Finding back doors and secret openings to the system
- Pretending to be an accepted component for crossing system boundaries
- Steeling codes and keys to get past system gateways
- Breaking down system gateways without setting off alarms
- Making an opening in the system boundaries without detection
- Hiding in elements that cross the system boundaries.

Once in the system and hiding among the parts, the infiltrator can then find the best place to affect parts and have the negative effects proliferate across associations. Unlike a crippling strike, an infiltrator can sometimes find a point of vulnerability that is not protected or find a process that can be manipulated. Infiltrators can work in secrecy for long periods of time to alter system activities, so that a final internal attack will do immense damage and infiltrators can alter system activities to make it more vulnerable external attacks. For example, an infiltrator can open the gates for an invading army to march in.

Bacteria and viruses are inherent infiltrators in nature, and they have killed many animal and plant systems since primordial times. For animals, some will enter the host across mucus membranes, others will enter through ingestion, and some must enter through wounds. Once inside the host, bacteria will generally either consume the host system from within or evade the immune systems through some level of symbiotic relationship with the host. Some weakened host animals can continuously live with a bacterial infection and spread that infection to others. Viruses, as described in Chap. 2 of this book, can spread rapidly across a host system and attack critical organs to cause death. Alternatively, a virus can also hide in the cells of the body for years in a semidormant state and become activated when the body is in a more vulnerable state.

In cyberspace, infiltrators include software viruses that embed into codes, computer worms that hide in memory, and hackers that create secret links to the outside. While the intent of these infiltrators can be complex, many have the ability to damage and destroy computer systems as well as cripple networks. Through the mechanism of computer control, these attackers can further cause the failure of machines and societal infrastructure elements. In the most extreme cases, the usurping of controls for advanced weapon systems and military sensors can cause false attacks or false decisions to attack based on misleading information. However, cyber security experts are perhaps most concerned about attacks upon the global financial system, which now depends completely on computer networks for the transfer of funds and accounting of assets. Once a cyber attacker has covertly broken through network defenses such as password gates, encrypted communications, and activities monitoring, an attack that could bring down the financial systems could either be a crippling strike at a financial hub with proliferated impact or a broad strike at a vulnerability that is common across many network computers. Either way, the mechanism of failure starts with cyber infiltration, and economic failure can be the disruption of economic processes through computer dependency, corruption of processes through compromised financial data, and destruction of economic processes through induced massive economic losses.

Finally, human infiltrators into a social or organizational system include spies, saboteurs, and terrorists. The techniques for these types of people are similar, but their missions are quite different. Spies gather and steal information to help opposing forces to defeat the system. Saboteurs destroy key system parts and links to reduce the system's ability to resist opposing forces. And terrorists seek to cause the most horrible events in the system to reduce the system's will to fight the opposition. However, human society is resilient, and the success of infiltrators often depends on collaboration with external forces. In World War II, for example, the resistance groups behind Axis lines blew up bridges and factories. However, the Allie forces had to quickly take advantage of the resulting system weaknesses, or adaptive response by the German system would have neutralized the effects. In the United States, the Al Qaeda attacks of September 11 caused horrible terror. However, the inability of Al Qaeda to exploit the social turmoil only led to the galvanization of the American will against terrorists.

3.1.5 Breakdown by Annihilation

A conflict is won when the opponent loses its will to fight or when the opponent loses its ability to fight. In the latter case, some systems consisting of flexible associations and minimally dependent parts can continue fighting even after great destruction of parts and disruption of links. The surviving parts can be reorganized to sustain vast percentages of system operations, and the system might be able to oppose attackers down to the last few working parts. Such a level of defiance calls for the total annihilation of the system. If the system does not have strong capabilities to reproduce parts and repair associations, then the attacks to achieve annihilation can be steady overtime. However, if the targeted system can reproduce and repair, then the annihilation attacks must be rapid enough to outpace the rate of restoration. Some systems are so resilient that even a few surviving parts can allow for a gradual buildup and the return of the system. Therefore, the comprehensiveness of the attack may be critical in achieving lasting system failure.

Within the human body, most invading systems must be completely annihilated because few virus and bacteria know when to retreat and hide. The process of annihilation in the body is typically a competition between the search-and-destroy capability of the immune system and the growth rate of the invading system. Medicine can support the annihilation process by stimulating the immune system to act early, strengthening the immune system to act with more force, helping the immune system identify invaders, and directly attacking the invaders through toxins.

One can make the argument that viruses and worms in computers spreading across networks need to be annihilated similar to their organic counterparts because they will not relent on being a threat, and they will multiply. Beyond the standard approaches of scanning and containment, computers can be shut down, and networks can be isolated until infiltrating systems are completely annihilated. If computer codes and memory cannot be cleansed, then enter environments can be wiped out and rebuilt to get rid of threats. This presents the reality that annihilating attacks are often not subtle. In the effort to destroy infiltrators, invaders, and rebels, massive regions of one's own system may have to be damaged or destroyed in the process. When this argument is made in the context of human organizations, the consequences yield serious value issues. How much danger must a society or organization face in order for leaders to consider sacrificing their own people to annihilate the threat?

Currently, the most devastating form of physical annihilation that can be levied by man is nuclear weapons. While nuclear weapons have been deployed to destroy hardened targets such as underground bases during the Cold War, their primary functionality is as weapons of mass destruction. A city hit by a nuclear warhead is completely destroyed, and the region around the multiple mile blasted radius is contaminated with lethal radiation for decades. Nature has its own ways of causing massive annihilation across human society. Throughout history, cities have been destroyed by volcanoes, earthquakes, floods, hurricanes, and drought.

The mechanism of devastation is unique for each form of attack by nature, but the end result is the breakdown of the human system and the death of many in the impacted region.

3.1.6 Breakdown by Corruption

Stepping back from the images of massive devastation as a system breaks down, a system can alternatively be broken down simply by the parts and associations not working. A conflict approach for disabling the system is therefore to corrupt key parts and manipulate key associations. Corrupting people in a societal or organizational system is as simple as applying threats, bribes, blackmail, and propaganda. The challenge in corrupting people resides not in getting people's behaviors to shift but in figuring out the patterns of shift. People react to external influences all the time and systems composed of people continuously struggle with self-organization and self-adaptive characteristics of the population. To shape peoples' reactions to get the desired corruptive effects, keeping the change straightforward is critical to success. Has the attacker been able to cause people to do unintended things that harm the system? For example, a propaganda campaign ahead of an invading army to convince people in the defending system to not resist could have corruptive effects. A get-rich-quick pyramid scheme launched by opponents can cause massive investment losses, economic instability, and demoralization of the population. The withholding of a life-sustaining or life-saving resource, such as medicine for a deadly disease, could cause people to act irrationally to do more harm to the system. Systems led by political leaders are potentially easier to corrupt because even a few corrupted politicians can cause great harm to the system. To guard against this outcome, democracies must have strong legal structures to discourage this form of vulnerability.

Corrupting computer systems is different than corrupting human components because the parts and associations are far less self-organizing and definitely not self-aware. Yet the complexity of these systems also yields vulnerabilities to corruption in the course of attack. Unlike infiltrating attackers that seek to take control of key computer functions or directly disrupt computer operations, corruption of a computer system involves changing the codes and data that enable computer operations. This change can be done by infiltrators. However, it can also be done by corrupting codes while they are being written and corrupting data before they are received by computer databases. In global computer networks, keeping track of where millions of lines of code have been written and who has written them can be challenging. Just validating codes through operational testing/scanning may not detect all corrupting content. Code validation efforts can identify intentionally buried elements. However, even a small vulnerability introduced into the system can be exploited by infiltrators to take down the system.

While the origins of code vulnerabilities are difficult to find, code vulnerabilities, such as the Heartbleed security bug in the implementation of the Transport Layer

Security (TLS) protocol using the OpenSSL cryptography library, has shown that systems can potentially break down due to code that does not operate according to purpose [7]. The intentional corruption of data involves creating false data and missing data that are not detected by the system and are then used by the computer system to yield negative operational outcomes. Data can also be considered as corrupted or compromised when their effectiveness is lost due to theft. For example, when millions of credit card numbers and associated personal information have been stolen, those numbers can be considered corrupted, even though the numbers have not been changed. This latter mechanism of corruption is common today due to criminal actions. However, it is unclear as to whether computer networks running global processes, such as international banking, have undergone concerted attacks by opponents seeking to break down the entire system.

Human and computer systems are vulnerable to corruption because they are complex. Complexity yields many places for corruption to occur, and complexity can amplify the effects of corruption. Simple systems in contrast can often be corrupted by changing just one behavioral driver. For example, if one chemical will interfere with the behaviors of insects, then that chemical can be used to corrupt and defeat insect colonies. If a mechanistic system uses parts that can be affected during manufacturing, then covert tampering might cause the parts to fail prematurely and unexpectedly during operations. The tampering of manufacturing processes can be through slight changes in the procedures. Other processes in organizations can also be corrupted through tampering with procedures, thus yielding physical security vulnerabilities, subtle but proliferated financial errors, disaster causing operational errors, and other damages. The potentials and consequences of corruption are very system specific. Unlike other failure mechanisms in conflict, corruption is not just about failed defenses and failed parts. It is also about how inherently unstable a system can be and how conflict is just a tipping force for systems to collapse.

3.1.7 Breakdown by Exhaustion

The last mechanism of system failure in conflict that we will explore is based on the reality that systems need resources and energy to operate. Even when all the parts and associations are intact, the depletion of a critical material or form of energy will break down the system. In the case of the human body, for example, the cutting off of air, water, or food will kill the body, as discussed earlier. In the case of human society, the depletion of fossil fuel will drastically reduce activities. In the case of computer hardware, the worldwide depletion of gallium, a rare earth metal, will hinder the availability of certain integrate circuits. And, in the case of the economic system, the depletion of cash flow will have catastrophic consequences.

During the height of the Cold War, the Soviet Union was trying to match the military manufacturing capacity of the United States, and the United States pushed the Soviet Union to the point of national exhaustion. Then, the Soviet Union literally broke apart as all the resources and energy were wasted on building war machines and as the Russian people lived in poverty. The idea of exhausting the enemy extends to the dawn of siege warfare where the invading army waited for those inside castles to suffer from dwindling supplies. In opposing invading armies, on the other hand, the defenders can burn their own crops and homes during retreat to deny replenishing resources for the enemy. This strategy was implemented by Russia during the invasion of the French army in June 1812. Unable to sustain his army of over 680,000 men in Russia, the retreat of Napoleon and the French army across the Russian winter admits continuous engagements by Russian forces led to the loss of over 500,000 men [8]. Thus, exhaustion in conflict can be an extremely brutal form of system failure because the system can no longer fight back.

At this point, I will attempt to further explain the value of taking this broad crosscutting look at systems breakdown due to conflicts. I understand that researchers with data on specific system failure events will want to focus on the data and the models using the data. However, even immense amounts of data can have missing elements, and even obvious models can have mistaken assumptions. To reduce the chance of being trapped by conventional research approaches, stove-piped thinking, and insular research communities, those studying the failure of systems in conflict should consider all the ways of breaking systems through conflict to determine the appropriate mechanism or combination of mechanisms that best reflect the situation. Then, our understanding of how systems formed, as presented in Chap. 2, can be used to ensure the comprehensiveness of the model. This integration of system breakdown mechanisms and system formation charac-teristics can be particularly useful in studying failure events with limited data or in projecting how failures can be achieved in future conflicts. Clearly, I am an advocate for exploring the problem space in a broad unconstrained manner and then converging upon the most likely behavioral path. On the matter of war, I view it as a barbaric flaw in human nature, and I, in no way, want to help people better understand how to inflict destruction upon others. However, wars may be necessary to stop human atrocities and to prevent greater loss of life. Also, understanding adversary approaches can help prevent attacks of mass destruction from succeeding. It is in this context that I present the failure mechanisms of human societies and human organizations. This respect for life will be maintained as I press forward to discuss other methods of system breakdown beyond conflicts.

3.2 Growth: The Unintended Damages

As conflict is the most obvious method of causing system failures, growth might be the most unsuspected method of inducing system breakdown. We have explored growth during system formation as the expansion of the structure with more parts and associations. Further, we looked at growth in connection with the types of system structures and at the impact of growth on system boundaries. Growth can be a part of normal system operations or it can start with system adaptation. System growth can cause evolutions in structures and behaviors. And reproduction is a way of growth that is in contrast to increasing system size and complexity. What I did not do in Chap. 2, however, is to single out growth as a top-level characteristic of system formation.

This is because growth is not essential in all systems. A system can be built and remain in a steady state until it fails or is retired, and a system can integrate with other systems in lieu of internal growth. Growth is, thus, mainly an organic process associated with biological entities, some human organizations and groups, human-driven information networks, and systems mimicking organic behaviors such as self-replicating machines and computer codes. Information technology and systems are remarkably able to sustain growth. While the installation of hardware and fiber optics are just processes in building the infrastructure, the logical networks that ride upon this infrastructure and the generation of data within this infrastructure can be highly organic in characteristics. Information systems have already gone through many factors of increase in capabilities over the past 40 years, and there appears to be no limit in their growth potential. Thus, in a section on system breakdown in growth, information systems and networks will be quite absent from the discussion. Instead, human organizational systems that depend on information can easily fail on growth, and there are more than enough examples of such failures to keep this section interesting.

From a system-definition perspective, growth is a path of change from one system state to progressively expanded system states. The expansion can be in the number of parts, the number of associations, the size of the parts and links, the complexity of the structure, the size/nature of the boundaries, and level of inter-actions. Growth can be identified as a unique dynamic path within a system that is a part of the overall dynamics for the system. Along this path, growth can have velocity vectors and varying rates of acceleration that yield a pattern.

The patterns of system growth can be divided into four general categories, as shown in Fig. 3.3. Steady growth, Category 1, reflects a constant velocity vector and is typically driven by a growth mechanism that expands the system at a linear rate. For example, a company might have a target of increasing revenues by 10 % each year and a growth mechanism where new employees are hired and integrated into corporate processes at a similar rate. Steady growth can still create stresses that break down systems. Thus, a system can grow and suddenly face a halt in growth. Some systems have self-limiting capabilities to prevent growth at harmful rates and to stop growth when early system damages are detected.

Regulated growth, Category 2, has a path where the system reduces the velocity through a decelerating mechanism as the growth approaches a limit. For example, the growth of cells in the human body will slow as the body approaches adult size, the population growth of animal groups might slow as food supply declines, and regulators might try to slow down economic growth if inflation is escalating out of control. Regulated growth is the pattern that is least likely to cause system breakdown unless the limits are set incorrectly or the system is actually stabilized by growth. For example, during aggressive economic growth, in which a failure mechanism has already formed, the buyers and sellers might all be so captured by the momentum of growth that the failure mechanism will get worst but effects of

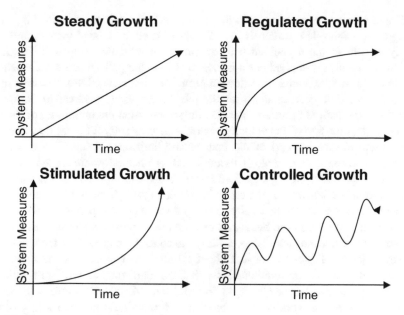

Fig. 3.3 Common growth patterns

failure are delayed. Such effects are then realized when the growth starts to slow down. In the great recession that started in 2008, the risky loans that caused bad debt to accumulate in complex bank portfolios had been growing for years [9]. However, the economic system kept pressing forward until home sales started to slow in part due to unsustainable annual price increases. As home prices went through an adjustment, the realization of the failure mechanism almost brought down the entire United States and world economy.

Some growth paths do not even attempt to find stability. In stimulated growth, Category 3, an acceleration mechanism causes the velocity of growth to continuously increase. The mitosis reproductive cycle of bacteria is a nonlinear accelerating mechanism, as it doubles the population every generation. Viruses reproduce hundreds and thousands of copies through each invaded cell to yield extremely nonlinear growth rates. Even the growth rate of human societies and organizations can be nonlinear if there are stimulating factors. Such factors include abundant access to food and resources, so that each generation can support many children, and accelerating market demands for company products that require matching company growth. Nonlinear growth is about the acceleration, and the growth generally stops when the system is broken or naturally slows down after the accelerating force is cut off. In the case of a bacteria population, growth is halted when the food supply runs out. Then, a massive population die-off typically occurs. The degree of system breakdown in stimulated growth is, therefore, often associated with the magnitude of the acceleration. The force that stops high acceleration might break the system into a million pieces.

The importance of shaping growth curves leads to the idea of controlling rates, Category 4. Controlled growth is the attempt to keep stimulated growth and even steady growth within a planned path or range. For stimulated growth, the deceleration mechanism is applied periodically to change the path of growth. Depending on the cycle of acceleration and deceleration, the path can exhibit minor to major oscillatory motion. For steady growth, the decelerating mechanism is applied throughout the path to cancel out accelerating forces or at the beginning of the path to change the direction of the velocity vector. This is slightly different than applying deceleration toward the end of the path as the limit approaches. The process of control can cause a great deal of dynamic stress on the system in the course of following a path and hitting objective states. Thus, the objective states might not cause system breakdown but the effort to get there will. Control is a challenge for human organizations with the ability to apply force at control points but without the ability to fully deal with the behaviors of human components. Companies can hire people, fire people, and establish procedures to control the growth of the workforce. However, the workforce is not just about number of workers. As the number of workers increase, ad hoc relationships will form, performance will shift, attitudes will change, and worker needs might rise and fall. Management efforts to continuously right size the workforce might further worsen attitudes and lower performance. Thus, in the attempt to control the workforce, management could drive the company processes toward breakdown.

I have suggested that the paths and dynamics of growth can lead to systems breakdown. However, the actual process of breaking down must have causes and mechanisms, and these will be different than the mechanisms we explored for breakdown due to conflicts. I have thought of a few ways for growth to breakdown systems, as shown in Fig. 3.4.

Once thing that is apparent between the mechanisms for breakdown in growth and the mechanisms for failure due to conflict is that breaking down in growth does not have to involve the corruption or destruction of parts. In systems growth, it is parts and not external adversaries that contribute to breakdown. This is explained as we explore the following ways and mechanisms of breakdown.

3.2.1   Breakdown Due to Too Many Parts

A system is defined by parts working together. This implies that the correct number and types of parts must be positioned and associated in the correct ways within the system. If the growth of parts in the system exceeds the system's ability to utilize those parts for structural growth, then the excess parts might become a hindrance to the operations of the system. The mechanism of system breakdown is, therefore, the unregulated or poorly regulated growth of parts in the system. Examples of this situation include human societal systems where there are no incentives and methods to curb the number of children per family. Thus, birthrate within the population depends on the individual attitudes of the people and religious views. The system's

Fig. 3.4 Ways growth can break down a system

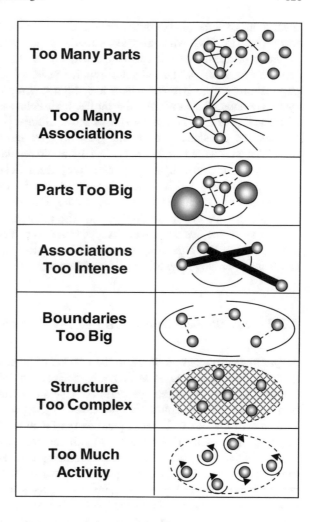

approach for handling the resulting population growth has typically been to grow the infrastructure and economy to match the needs of greater population. This is in contrast to bacteria and other animal populations that experience massive population die-offs when resources dwindle. With the world population surpassing seven billion people, the ability of societies across the world to sustain their growing population has come into serious question. Some societies, such as that of China, are sacrificing their environment to provide a better material quality of living for their massive population. China has also struggled to enforce a one child per family policy to reverse population growth. Other societies are plagued by civil unrest due to high unemployment levels, poverty, and lack of safety.

High-density population areas with weakened societal infrastructure have contributed to the rise of epidemics, starvation, crime, and illiteracy. Under such conditions, societal systems have collapsed. At times, the collapse ends in revolution, with the people overthrowing the government, as in the case of France from 1789 to 1799 with crowded cities and detached nobility [10]. However, systemic problems are difficult to correct, and building better societal structures from the remnants of the old often require years of struggle. In some cases, the population grew to a point where a sudden environmental shift quickly led to system collapse. This occurred during the Irish Potato Famine from 1845 to 1852, where a disease wiped out potato crops across Europe [11]. Prior to the strike of the disease, the population of Ireland grew in size and dependency to the potato crop. The famine led to the starvation of approximately one million people and another one million people emigrated away from Ireland. The escape of system parts as processes fail and boundaries dissolve is a common behavior of this mechanism of system breakdown. As noted, the parts do not have to die or be destroyed for the system to collapse. In the effort to escape, parts critical to the operations of the system might disappear and too many parts might leave. The system with self-proliferating parts might end up with too few parts. Thus, the challenge is on how to prevent self-proliferation from defeating the system.

3.2.2   Breakdown Due to Too Many Associations

A system does not have to increase the number of parts to grow. The growth can be in the number and type of associations established within the system as well as between the system and eternal entities. A certain number and pattern of associations are required to capture the processes of parts working together. As the performance of the system improves and as the function of the system increases, one might expect to see an increase in the number and type of associations. The transition of a system to being more self-organizing and self-adaptive should result in more associations. However, the propensity for parts in a system to self-establish associations could lead to a vast number of excess associations that distracts the system core processes. This distraction can become a dynamic hindrance when the excess associations are between system parts and elements of the environment. For example, workers in a company will go home and have separate lives and associations. Then, when the workers bring those associations into the corporate system, corporate processes can get disrupted or delayed. Workers might use company relationships to advance side businesses. Workers might use company time and communication resources to work out personal relationship issues. And workers might betray company secrets to competitor companies. These collateral associations can grow to dominate the affairs of the company.

As collateral associations grow out of control, the mechanism of system breakdown is by parts becoming ineffective at working together. Parts governed by many associations cannot devote enough energy to the associations that sustain the operations of the system. All the parts in the system may still be present, and all the critical process links may still be in place. However, the dynamic activities of system will encounter problems if the processes are carried out slower, with greater errors, and/or with reduced capacity. Any one of these performance issues can cause the system to be defeated in competition or to be penalized for not meeting standards. In some cases, the damages due to performance errors can be so great that the system will collapse in the wake. To guard against distracting associations growing among system parts with critical responsibilities, some parts, such as workers handling nuclear weapons, may have to be periodically evaluated to determine readiness. The complexity of system parts and their ability to self-organize will yield insights on their excess commitments. The challenge is on how to prevent self-organization activities from harming the system.

3.2.3 Breakdown Due to Parts Being Too Big

Beyond parts self-proliferating, parts in systems can also grow based on the characteristics discussed in Sect. 2.1. Parts can increase in size relative to their reference frame. Parts can have more complex internal structures. Parts can increase their material, energy, and information content. And parts can increase their surface features and dynamic behaviors. The first question to ask, when parts in a system start growing on their own, is whether the parts can still sustain the associations and processes for which they are responsible. The second question is how well they can perform their core functions given the demands of growth and the distractions of growth. If associations and performance are sustained, then the system might not break immediately. However, the system also might not realize the harm and risks posed by growing parts until breakdown is unavoidable.

In the human organizations, people can grow in knowledge, physical strength, skills, and credentials. This often helps the performance of the organization, as the execution of processes can be enhanced through relevant growth. However, what if the growth is not needed by a person's role in the organization? If a doctor has gained the skills to become a world-class golfer, how will the expanded capabilities of that person impact the medical system? One system part that has grown beyond its role can be replaced. However, what if all the doctors want to become professional golfers? Then, the system will have a problem. Though the example is extreme, the loss of experienced workers due to personal growth has troubled companies and caused performance problems. Even when workers merely believe that they have additional value, their demands upon a company individually and collectively could yield large financial losses.

In economic systems, the growth within parts could mean the increase of productivity for economic units. Factories can grow to have higher rates of output. Farms can grow to yield more crops. And schools can grow to provide more professionals. However, if this growth does not stimulate demand, then the surplus resulting from growth will cause deflation and economic instability. As prices fall, people with real needs might hold back on purchases, thus slowing down the economic system. As values decrease, such as in the price of homes, the significance of debt becomes more real. Thus, imbalanced growth can cause economic slowdown and even system breakdown as other forces take hold. These examples show that this mechanism of growth failure is connected to the adaptiveness of system parts. The challenge is on how to keep self-adaptation focused on the needs of the system and not the conflicting needs of individual parts.

3.2.4 Breakdown Due to Associations Being Too Intense

Associations between system parts can grow as the parts themselves can grow. An association can increase in carrying capacity to pass more information, forces, energy, and substances between parts at any point in time. An association can increase in throughput to pass at a faster rate. And an association can grow to pass broader ranges of information, forces, energy, and substances. Usually, one would expect that growth in an association is stimulated by the growth or changes in parts. However, the interactions between steady-state parts might not always be stable, and the associations can grow on their own. This growth might cause too much flow between the parts resulting in damaged parts and perhaps the detachment of the link. Even when the growth is stimulated by one growing part, the other end of the link might be too overwhelming for the attached part.

An example of an association growth between people is normal communications within the organization escalating into passionate shouting matches over organizational process related issues. Attempts by the person at each end of the link at working together to achieve system results could actually fuel the intensity of the association. At some point, we can imagine work halting as the act of verbal exchanges takes over. At some point, we can imagine the link completely breaking or the intensity spreading to other people in the organization. This growth of association intensity is different than the growth in number of associations, as discussed earlier. Intense associations are still structural connections sustaining the processes of the system. Thus, they are not something that the system can simply cut out and isolate. Instead, the intensity must be managed before the breakdown of associations collapses the system.

Another example of growing associations is the workload of people within a process escalating as assignments are passed from person to person. As one person increases the rate in which he or she completes assignments, that person then drives the performance expectation for others. As different people increase their rates, the total process or system will start to spin faster. Initially, this might be a welcomed change. However, the system must be brought to some stable state where the growth slows down. Otherwise, the process will spin to the point of failure. Unlike the growth of system parts, the growth of system associations can be stopped by breaking the links. After the parts have stabilized, links can perhaps be reset to proper levels, assuming that the system processes can withstand the disruption. Going back to the communications example, sometimes it is better to simply send people to different rooms and let them cool down. These examples show that this mechanism of breakdown due to growth is connected to the flexibility of system associations. The challenge is on how to keep flexibility within the range that is tolerated by system parts.

3.2.5 Breakdown Due to Boundaries Being Too Big

The boundaries of systems with dynamic and adapting parts can change. Even when the boundaries are fixed, the parts can push out the boundaries to other fixed states. When the boundaries are more elastic, the parts can stretch the perimeters to elastic limits. Thus, system boundaries can grow, even though the number of parts and associations has stayed the same. The growth of boundaries as described is different than external forces changing boundaries by design. Builders sometimes physically pull apart the boundaries of systems, and control authorities sometimes refine the rules for boundaries such as that of a political state or census district. Systems breakdown due to these actions are, therefore, failures of design and not failures of growth. Growing boundaries refers to systems that are not centrally controlled, such as that of a naturally formed human society. The boundary of such a society is typically based on how far away people are willing to settle and how much territory people want to claim.

Continuing with the society example, the stable growth of the boundary is dependent on the expansion of the societal infrastructure and the growth of the population. However, history has shown that this is not always the case. Population has sometimes spread out in search of opportunities even as associations break and processes weaken. Armies are sent out to conquer more territory than they can control. Further, social chaos can also cause boundaries to stretch. One of history's most extreme cases of boundary extension was the Roman Empire [12]. Although there are many theories regarding how and why the Roman Empire fell, we identify several facts: (1) the Roman legions spread out across Europe beyond the natural boundaries of Roman society; (2) the conquered people of European lands were not initially integrated into Roman society as citizens; and (3) Roman citizens largely did not emigrate across Europe to extend the Roman system into local social systems. Thus, from a system boundary growth perspective, we must think about the resulting vulnerabilities that were introduced. As boundaries faltered from enemy invasions after the third century AD, Roman territories were quickly reverted to local cultures with pockets of Roman influence. As the Roman armies withdrew, only the Roman roads and ruins remain. The Mongol Empire of the thirteenth century was perhaps another example of rapid growth and system instability. With a very low population, the Mongol tribes conquered most of Asia and much of Europe in the span of a hundred years. However, the Mongolian people ruling over the massive population of China were, in the end, either driven back to the Mongolian homeland or absorbed into the Chinese system [13].

I certainly do not wish to trivialize history and debate the theories of historical events. However, I do believe that the study of systems can offer new ways to think about history. The lesson from history may be that system growth can be easy but maintaining system stability is hard. The growth of economic boundaries can also be studied as an example of system breakdown. It is easy in modern times, with worldwide communications and travel, to form ever-larger trade unions and establish common currency. However, just because one has created a larger system through extending the boundary does not mean that all the parts will grow to match the demands of the system. The new boundary might instead allow parts to be stretched beyond their performance limits. Once again, system theories are far more complex, and even a systems model of the growth vulnerabilities that I have highlighted would be quite elaborate. So these examples are merely to show that the mechanism of failure due to growing boundaries is a system concern connected to the flexibility of the boundaries. The challenge is either on how to grow the boundaries in a stable manner along with the growth of the system or on how to grow the system structure to keep up with expanding boundaries.

3.2.6　Breakdown Due to Structure Being Too Complex

Having many parts and many associations can cause system breakdown. However, the breakdown may alternatively be due to the complexity of the system structure. Structural complexity depends on the number as well as the positions and arrangement of the parts and associations. Just like a box of Lego blocks, pieces for building the system can be fitted together in many ways. A thousand pieces can used to form a simple cube, but the same thousand pieces can be used to create the objects of our imagination. The difference between building Lego structures and real-world structures growing in complexity is that the Lego builder typically adjusts the design and block arrangement techniques through feeling out the stability of the structure. Such an adaptive control of structural expansion might not always be present in self-promoted real-world structural growth. An example of structural complexity that is self-generated within a system is the growth in the complexity for social networks. The people on a social network can be a set number, and the links enabling people to interact with one another can also be mapped as an extension of the network infrastructure. However, the actual organization of groups and subgroups with overlaps, access barriers, and extended privileges can grow progressively more complex. Who is in whose friendship circle? Who is blocked from seeing people's shared information? Who can gain access through mutual friends? These are all ways to probe the complexity of a self-grown complex structure. The question is whether such a growth can lead to structural collapse.

In the case of a social network, the collapse comes through the mechanism of system parts becoming detached from the structure they have formed. If the heart of the structure is communications, the excessive communication dynamics can turn constructive or entertaining dialogue into mind-numbing chatter. Three people sharing information is indeed a dialogue. However, when 10,000 people are trying to talk, all we hear might be noise. As people become overwhelmed by social networks, activities might decrease and networks might decline. To prevent system breakdown from structural complex, strategies have been knowingly and perhaps unknowingly implemented to include: (1) limiting the pathways of communications, such as criteria for allowing people to connect; (2) limiting the scope of content that can be communicated, such as set fields of information; and (3) limiting the purpose of the network, such as for a single shared interest.

Another example of excessive structural complexity might be the financial instruments invented by bankers and corporations to organize debts, control currency flow, shift the location of documented profits, distribute losses, and value assets. Over many decades, some of the smartest minds in management, economics, and finance have devoted themselves to creating this complexity to benefit the rich. In the end, these brilliant and heavily educated minds apparently lost track of the true risks in the dynamics of such financial instruments as demonstrated in the Great Recession that started in 2008 [14]. As financial loss spread across the economic structure without anyone fully understanding how the complexity has hidden the associations or is amplifying the effects, the economic system, which has grown for years, was going to collapse without government intervention. These examples show that system structural growth can lead to a mechanism of failure, where parts become detached from system processes and system processes can spin out of control within the structure to further breakdown or eject parts. The challenge when structural growth is placed in the hands of many participants is on how to steer growth away from risks while retaining the benefits of growth.

3.2.7 Breakdown Due to Too Much Activity

The last breakdown mechanism related to systems growth that I wish to explore is driven by the one type of growth that is not centered on human and organic systems. As all systems are defined by their dynamics, many systems can grow in dynamic properties, even as the parts and associations remain the same. Very rigid systems, such as mechanical devices, still sometimes become more dynamic over time. Parts can spin faster after operations, engines can burn hotter over time, and the system can become more maneuverable based on increased environmental information. In human and biological systems, the self-adaptiveness of parts can cause changes or growth in dynamic properties. In human organizations, the workers can become more integrated as a team to increase output. In human networks, participants can learn more about one another to increase the dynamic range of communications content. In animal herds, the herd leaders can become more responsive to the dangers of the environment.

As first glance, growth in system dynamics is not always dangerous. In the days before precision manufacturing, machine parts were expected to wear in and work better after use. For engines that were initially not using fuel efficiently, the efficiency grew as the valves and pistons become more settled in. Even today, for systems that are initially uncertain about how to navigate, accumulating environmental data can lead to a growth in maneuvering capability. These types of dynamic changes are, however, system growth by design. In general, manufactured systems will encounter integrity issues if dynamics exceed designed performance ranges. Even when the dynamics appear to be able to grow, the speed of the growth could induce stresses that shake apart the structure, and maneuvering of the system could place it in environmental conditions that the system cannot handle. For human and biological systems, dynamic increases due to integration, mutual awareness, and situational learning can also be beneficial. Further, there are often no defining performance ranges. Performance, however, is a multifaceted set of metrics. If all the people figure out how to work twice as fast, would the people also become exhausted before planned rest periods? If everyone communicates based on deeper interpersonal knowledge, would the communications start to hurt people because of selfishness and jealousies inherent in human nature? If herds of animal start to learn how to navigate through the environment, would a sudden change in the environment place the entire herd in greater danger? These questions reveal that, while system dynamics can grow, the growth can bring about additional risks over steady-state operations. In fact, even when the change in dynamics is not growth but merely due to ad hoc maneuvers, there is a higher risk of accidents. Plane accidents occur more frequently during takeoff and landing, and car accidents occur more frequently during sudden stops and lane changes.

I started this section by suggesting that growth may be the most unsuspected method of system breakdown. However, once we realize the risks in growth, the mechanisms of system breakdown might be easier to model than breakdown due to conflicts. This is because growth is mainly about modeling the system and the conditions that stimulates and sustains growth. Conflict, on the other hand, requires the modeling of interacting systems, strategies and tactics, decision processes, terrains of combat, effects of damages, and the chaos of war. To model the system for studying growth, we can rely upon the frameworks established in the sections on how systems form. Once a system model has been constructed, the dynamic ranges of growth, sources for growth instabilities, and scenarios for system breakdown can be explored for a specific system.

Though growth embodies risks, I, by no means, want to advocate for halting growth, for some will say that anything that is not growing is in the process of dying. Growth failures are often due to extremes and so steering growth is about not crossing the line where many new parts becomes too many new many parts and many new associations become too many associations. If this line is too difficult to identify, then the system should perhaps be pulled back in the trade-off between growth and instability. We have seen attempts at trying to prevent economies from overheating, and we have seen countries, such as China, attempt to curb population growth. In pulling back the system, we need to understand the line where the system will start to decay. This lower line will be explored next.

3.3   Decay: The Unavoidable Breakdown

If systems are not destroyed by conflict and if systems do not fall part during growth, then they might operate for long and productive lifecycles. However, all physical systems will eventually breakdown, and all information systems that have physical components, such as computer disks and chips, will need to replace the physical components over time. Even the data that supports information processes will loose quality over extended periods of use. This is because system dynamics cause parts to change, components to wear out, and data to accumulate bit errors. Essentially, the process of decay in system dynamics is unavoidable.

Decay can be slowed down through better systems design, stronger construction of parts, and more carefully executed system operations. However, unless decay is countered with the repairing or replacement of parts, a system will eventually have to cease operations. There are no perpetual systems in the real world, but some systems may appear perpetual because all the parts have been replaced overtime. The total replacement of system parts is simply the process of building a new system upon the decaying structure of the old. With such a process, airplanes a hundred years old can be restored to flight, and software filled with corrupted codes can be reloaded from a stored master code set. In societal systems, human beings age and die. However, the cycle of birth and learning sustains and advances society. After generations, we can ask whether a society is the same system or a system that has been reborn many times. Certainly, cities have been excavated to reveal layers of past cities beneath.

Before understanding how to rebuild decaying systems, we should first study the process of decay. This process is simplified into two portions, the type of decay experienced by the system and the dynamic path of decay experienced by the system. If a system can experience multiple types of decay and multiple paths of decay, then the potential combinations of types and paths will yield very complex patterns. To start with the types of decay, I wish to first suggest that decay is primarily related to system parts. The decay of parts can break down associations and the structure. However, if the breakdown of associations is not caused by the decay of parts, then a different method of system breakdown might be at work, or the breakdown of associations might be better defined as another characteristic of the system. Some associations, such as goods transport and Internet communications, might depend on physical elements such as roads and fiber optics. Such elements should be considered association-enabling parts, and these parts will decay like other parts. Thus, to focus on parts experiencing wear and tear during system operations, the types of decay, as shown in Fig. 3.5, are the part completely wearing down to the point of failure, the part malfunctioning relative to its normal operations, the part and associated parts becoming weakened/vulnerable, and the part morphing into a new structural configuration.

Decay Type 1: The simplest form of decay is a part wearing down from dynamic activities until it either breaks down or stops working. For physical parts, the wear can come from surface contact, internal operations, energy transfers,

Fig. 3.5 Types of decay

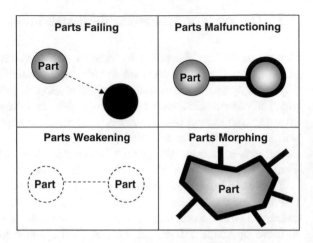

material exchanges, structural movements, and other external forces. For information-based parts, computer codes can become corrupt in memory, and data can accumulate errors every time an operation is conducted on the database. Faulty codes can cause computer systems to crash, and faulty data can cause erroneous system performance or process confusion. In economic systems, what wear down are the people driving the economy. What if too many professionals, such as doctors and engineers, want to retire? Simple decay should be easy to identify through parts examination or through following the source of broken links. At the source, the decayed parts might remain as husks or the remnants of parts might be completely absorbed into the system environment. Each case could uniquely impact the detection of failed parts.

Decay Type 2: A part might decay to a point where it malfunctions. Then, the decay might slow down because of the malfunction to allow the part to remain in the system for extended periods of time. Since the malfunctioning part is still connected to the system structure and impacts system processes, the effects on the system is a trade-off between those system processes that are still minimally supported by the part and those system processes that are adversely altered by the nature of the malfunction. For example, a malfunctioning computer guidance module in a vehicle might still allow the operator to use computer guidance. However, the computer guidance might be wrong 25 % of the time, thus causing the operator to continuously assess the effectiveness of the system part. The debate regarding malfunctioning parts then revolves around whether to rip out the parts and when the parts will fail completely. The ease of repairing or replacing the parts is a factor in this debate. If not dealt with, the seriousness of the malfunction can progress over time. Early malfunctions in system parts due to decay might be so minor that they are not even detected. In fact, if minor malfunctions can be detected early and corrected, then the concern of total system breakdown might be a moot point.

Decay Type 3: The decay in systems might be somewhat uniform across many parts instead of affecting select parts more than others. In such a case, vast groups of parts and links weaken together before any specific part breaks down or malfunctions. The weakening of parts might have external indicators, or the weaknesses could progress to very advanced stages without changes in system dynamics. The danger in the undetected or undetectable progression of weaknesses is that a mass failure of parts can occur after system dynamics reach a certain point or encounter some external trigger. For example, a societal system with a drastic decline in birthrate might see its workforce functioning but aging for decades. Then, once the bulk of the aging workforce reaches retirement age, the society might experience dramatic upheavals. Even worse, new pollutants/toxins in the environment weaken the cells of people over time, but such weaknesses might not be realized until cancer starts to increase across the population. Understanding the long-range consequences of uniform decay is therefore critical in responding to the weakening of parts.

Decay Type 4: The final way in which a decaying part can change is to become something else. The decay process can cause structures within a part to weaken and morph into another configuration. The new structural configuration can yield new features and behavioral properties with the part. These behaviors can then proliferate across linkages to impact the total processes of the system. In the section on conflict, I suggested that cancer is a revolution within the body. However, cells can decay into a tumor configuration instead of becoming cancer cells or prior to becoming cancer cells. A tumor cell is still a part of the body, but its structure is heavily deformed. The structural transformation within the tumor can be brought about by physical trauma, chemicals, or simply age. As the defective tumor cells still operate and multiply, the growing tumor can affect body functions, particularly if the lump is within a vital organ. Modern surgical techniques have reduced the threat of tumors. However, it is still a good example of the potential consequences of decay.

By now, it is apparent that the mechanisms of decay are simpler than those from other methods of system breakdown. Instead, the paths of decay are what are more complex. This is because decay is generally a longer process when compared with conflict and growth. The potential length of the process then allows for variations and patterns, as shown in Fig. 3.6 and discussed afterward.

3.3.1   Subtle Decay Path Until Tipping Point

The Tipping Point is a well-known book by Malcolm Gladwell that provides evidence showing systems often decay to a point, which is unnoticed, and then suddenly collapse [15]. The decay can be by any of the mechanisms described. In the case of failing parts, the system could operate and adapt to the failures until one part critical to the adapting system processes fails. In the case of parts malfunctioning, the system could operate with malfunctions until a point where the operations

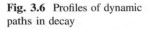
Fig. 3.6 Profiles of dynamic paths in decay

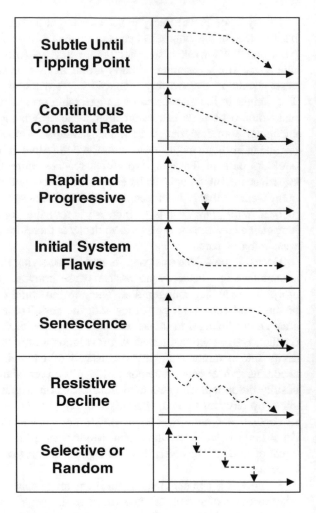

cannot handle the malfunctions. We have already noted that the uniform weakening of parts can lead to a massive sudden collapse. Finally, the system might even be able to operate with transformed parts until the tolerance suddenly snaps. This path of systems decay and breakdown is similar to the failure of material under stress and strain. Essentially, the system, like the material, will resist failure until the very end, and then the structure snaps, the boundaries fall, and the dynamics stop.

The challenge in studying this decay path is finding the tipping point because, until that point, there might be actions that can be taken to prevent systems breakdown. Once that point has been crossed, the argument is that collapse is virtually unavoidable. To find that point, one must understand the structural and behavioral properties of the system under stress, and this requires modeling, as discussed. Then, one must understand all the external and internal factors that could trigger the collapse. Conflict is clearly a factor and growth is another factor. Though seemingly inconsistent, a decaying system could attempt to grow and collapse in the process of growth. If so, we can ponder whether the breakdown is really caused by growth or decay. Beyond the obvious extreme forces, even small environmental changes that affect the final parts can cause a tipping point. The concern with global warming, for example, is whether there is a tipping point where global temperature rise suddenly triggers a massive worldwide climatic shift. Take the simple case of heating water, one degree below the boiling point and the water stays in the pot, and at the boiling point the water becomes vapor. The potential for tipping places great concern on any system with parts going through a slow decay process. Economic concerns and concerns about disturbing the system may require delays in repairing or replacing decayed parts. However, delays contributing to the tipping point will be disastrous.

3.3.2 Continuous Decay Path at Constant Rate

Some decay processes might not have tipping points. Instead, the cause of the decay is a constant activity, such as parts exposed to external stresses steadily fatiguing as the system moves. The exposed parts will, thus, decay at a linear rate, and the system will lose parts at a linear and observable rate. This path of system failure generally applies to systems without critically interdependent parts. Thus, the system can afford to lose parts while still remaining in operation. The structure of military systems anticipating high attrition often adopts this design principle. If successfully designed, the delaying of systems breakdown is then focused on slowing down the constant rate of decay. For example, in Napoleonic-era warfare where troops line up and time on the battlefield is measured in rounds of fire plus reload, the failure of the military system depends largely on how many troops are lost per round of fire. In this specific situation, system failure can be due to conflict mechanisms, as discussed, as well as a decay mechanism.

Ways to alter the constant rate of decay includes: (1) reducing the external forces, energies, or information that are wearing down system parts; (2) changing the properties of system parts to make them more resilient; (3) changing the position of system parts to reduce their exposure to decaying effects; (4) establishing barriers against elements that will decay parts; and (5) enacting a process to repair/replace decaying parts. All four types of parts decay discussed above can follow a constant rate of decay. However, the relationships in decay rates between the parts and the system could vary. For example, all the parts decaying in parallel will lead the system to break down at a similar rate. In contrast, when parts are decaying in overlapping sequences, the path of system breakdown can be slower with even some changes in rate along the way. Continuous decay in systems is generally more apparent than other decay paths. However, if the cause of the decay cannot be halted, then the system will break down.

3.3.3   Rapid and Progressive Decay Path

Some causes for the decay of parts have compounding effects. As the effects of decay accumulate in a part, the decay can accelerate. Within the system, conditions caused by parts failing can promote other parts to fail. The result is a nonlinear rapidly descending path of system breakdown as the decay progresses. Toward the end, the system essentially drops down to failure. An example of progressive decay is a compound that causes parts to degrade, such as a carcinogen attacking cells, and the progressive accumulation of the compound to accelerate the rate of decay. Along this path to system breakdown, every second matters in halting the accumulation of effects and slowing down the decay rate. This requires early detection and immediate intervention. Ironically, the initial rate of decay along this path might be slower than that of constant decay. Further, one might not even be able to tell the path until the acceleration picks up. If so, the window of opportunity to detect and act is further narrowed.

Ways to counter rapid and progressive decay includes: (1) finding and negating the cause of decay before the effects of decay escalates; (2) slowing or stopping the accumulation of decay causing elements; and (3) altering the vulnerability of parts to decay causing elements. What is not mentioned above is the approach of repairing or replacing decaying parts because it is unlikely that any response cycle can keep up with a nonlinear progression of system breakdown. Repairs and replacements can help, but it must work in conjunction with a way to slow the progression of decay.

3.3.4   Decay Path Based on Initial System Flaws

In the early days of manufacturing, where quality control was not perfect, a certain percentage of systems produced will have manufacturing defects. These defects will cause the system to fail early in the life cycle. Then, the number of system failures will drastically reduce and the remaining systems will generally operate well until the end of their lifecycle. Those flawed systems that have failed early were termed to have experienced "infant death," not necessarily a kind term [16]. With higher manufacturing precision and better prerelease testing, early life cycle system failures have been dramatically reduced in today's production systems. However, failures still occur particular with systems that are produced in low volume and systems that have individually tailored features. This is because the lack of a large number of samples could reduce our engineering understanding of unplanned and unanticipated decay processes. Even today, some satellites will fail early in orbit and some cars are still lemons. The challenge with satellites is that they operate in the harsh environment of space and limits on their weight discourage excessive redundancies, backups, and component hardening. In contrast, commercial aircraft are designed to handle forces and environmental conditions far beyond their intended operating conditions because human safety has been placed as a high priority. Returning to the term "infant death," natural systems also have flaws in the early-stage development and growth process. Even without flaws, the growth process of organic systems may yield greater vulnerabilities to decaying elements such as environmental toxins, diseases, and nutritional imbalances.

Like other paths of decay, early life cycle decay can embody any of the decay types discussed. However, the decay will center on flaws and vulnerabilities in the parts. For example, a part might have a material weakness that will cause it to break under stress, and a part might be prone to malfunctioning based on operational triggers. Hardware components of information systems can fail early in use and software components of information systems can fail early due to flawed installation and weaknesses in design and coding. Coding errors can allow malicious elements such as viruses, malware, and hackers to take down the system, either through a failure mechanism resulting from a conflict base or through a failure mechanism resulting from forced decay. The nonlinear system breakdown curve at the beginning of the life cycle is influenced by the nature of flaws and vulnerabilities as well as the stresses experienced by the system during initial operations. Some might argue that systems should be put through heavy operational stresses during initial use to discover those units that will fail early. This reduces the chance of surprise failures as the operational life stretches on. Alternatively, if system problems can be corrected through recalls, then manufacturing repair capacity can create more preferred paths of failures at end of life.

3.3.5 Decay Path of Senescence

The counterpart to early life cycle system failures is the end of life cycle failures. As a system ages, decay will accumulate from a variety of sources. In fact, it is almost not worth trying to figure out where all the factors affecting decay are coming from if the average system life span is of acceptable length. Instead, the focus should be directed at how the system breaks down in the end after a long and productive life. This path of decay typically reveals a long period of steady system operations, a progressing rate of failures in system parts, and finally the breakdown of the system. Once the effects of age start to take hold, the decline is quite similar to the rapid decay discussed earlier. In systems exposed to harsh environments, such as the human body, the senescence path is integrated with other paths of system breakdown such as wars, diseases, accidents, crimes, and exposure to harmful substances. We see death rates increase in the population for those after the age of 45, and then we see the rapid escalation of deaths for those after the age of 65. By the age of 80, the slope of system/body breakdown is practically vertical with few people living to 100. As the human system can repair and renew itself, there has been much research and debate regarding whether the human process of cellular decay is a feature of the system that can be reversed or delayed. Currently, some animals have longer life spans than human beings, and some humans have lived up to nearly 120 years of age.

From a systems perspective, the search for factors within the system that govern decay at the end of life is very different than searching for external elements that cause decay. At the beginning of life, the external causes and the internal conditions associated with system breakdown are integrally connected. At the end of life, after a system has dealt with external forces for a long time, what makes that system vulnerable to breakdown is more a characteristic of the system. Even if external elements have been weakening the system parts for years, what makes a part start to rapidly decay at a specific point is a property of the part and system. In the case of the cell, the search for factors connected with end of life decay has led to genetic states such as telomere shorting and questions such as whether genetic restorations can rejuvenate cellular activities [17]. For mechanical systems that are approaching the end of life, heroic lifesaving methods have extended operations. Such methods have been especially useful for satellites in orbit, as parts cannot be easily replaced and replacement systems might be delayed due to launch schedules. To explain, if aerospace engineers can figure out ways to modify satellite operations via ground-based controls to compensate for reduced performance and failed parts, than satellite continuity of operations gaps can be avoided and perhaps even the total cost of satellite ownership can be reduced. Just because systems must face senescence does not mean that death has to come at a specific time.

3.3.6   Decay Path of Resistive Decline

Attempts to counter the effects of decay lead to rather segmented or even oscillatory paths. Along such paths to system breakdown, decay will happen, and the system will try to either halt the progress of decay or repair the damages caused by decay. If the decay is linear but unceasing, then the resistance might achieve a balanced operational state or a much slower rate of decline. If the decay is nonlinear or if the decay causing elements also increase to oppose repairs, then the path of system decline could oscillate as it descends. I have suggested earlier that rapid and progressive decay is almost not worth countering if one cannot figure out the cause of the progression. However, aggressive efforts to resist such a progression could, nevertheless, result in an oscillatory decay path. This path is potentially important if every day or minute of life matters, but the effort to resist progressing decay can be increasingly traumatic to the system as the end of life approaches.

In the earlier section on conflict, I suggested that cancer is a cellular revolt within the body. As the conflict is lost and cancer cells spread rapidly across the body, cancer growth can be viewed as degrading organs along a progressive decay path. Unlike external elements driving decay, the self-proliferating nature of cancer makes it difficult to stop. Instead, efforts to locally remove cancer growths by surgery, kill cancer cells by chemicals or radiation, and supplement organ functions that are failing can delay the death of the body. At times, the patient will even feel better for short periods until the growth of cancer again dominates. Many terminally ill cancer patients have undergone the debate of whether to die while resisting cancer to the end or to accept the path of decline to embrace the fewer remaining days of life with more passion. The cost of resistance and the value of limited extensions to life are the governing factors in the shape of this decay path.

3.3.7 Selective or Random Decay Paths

As one might have guessed, decay is often studied through statistics. This is typically because the unique modeling of decay for every single part in a system is too daunting. If we increase our understanding of the mechanisms of decay, however, there is the potential of discovering behavioral patterns and metrics that reduce reliance on statistics. Whether by statistics or deterministic equations, the different paths of decay all assume a continuous or connected process. Thus, the decaying parts can be treated as a group, and one failed part has a relation to other failed parts. In some cases, the activities of decay are not connected, and the paths of decay will not be continuous curves. Each decay activity is then a selective event, or the occurrence of disconnected decay activities can be regarded as random for the purpose of studies. The problem with a path of shifts and changes is that understanding one segment of the path or one event does not necessarily allow us to understand the rest of the path. The path can be formed by many decay mechanisms interacting within a system at different times, and parts experiencing a variety of decaying effects can create uncertain outcomes for systems breakdown. This path of shifts and changes can still be modeled, but the model must have many agents and allow for segmented results for different periods of decay. If an accurate predictive model cannot be formulated, then the decay path will have to be tracked continuously, and any response to decay must be adaptive in real time.

As we finish this cursory look at the many interesting profiles of paths for system decay, one might ask why does decay need to be studied if a system has already achieved a productive life? Despite a successful life, the decay process might be too disruptive and perhaps painful in the case of human systems. The variability of the point where the system completely breaks down, such as a plane in midair, might yield too many dangers. And, the way breakdown occurs, such as an oil spill from a tanker, might cause collateral damages. If these outcomes can be predicted, the system should perhaps be either steered toward a less troubling decay path or retired before the negative effects start to emerge. Intentional termination of the system is one strategy for managing the end of life, but it should not be a step taken trivially. Even when the system life has been extended through heroic measures, the eventual breakdown can be chaotic and uncontrolled. In the case of human life, the last days of the terminally ill can be painful and tragic. Therefore, we can debate to what extremes should doctors extend life by a few days or a few weeks. Also in the case of human life, modern medicine can, in some cases, keep the precious body alive for a long time through machines, even when the mind is dead and the heart has broken down. At what point has the system of the body completely broken down and when should our machines stop sustaining the remaining living cells of the body? The human example highlights that even defining what is system death can be challenging.

3.4   Obsolescence: The Planned Breakdown

Since I have suggested the termination of systems before they breakdown in more damaging ways due to decay, the next question is whether there are other reasons to intentionally shut down a system. If so, then the process of system shutdown can be considered another method for how systems break but not necessarily for how systems fail. The traditional rationale for shutting down systems that are still fully operational has been that the systems have become obsolete. As Fig. 3.7 shows, there are many reasons for obsolescence. However, the conclusion is always that it is better to remove the system or replace the system. Thus, we will explore ways to breakdown the system after examining rationales.

I should note that the rationales for obsolescence apply to systems where a value judgment can be made. This is not the case for the physical system of the human body because human life cannot be measured by simple value metrics. The least capable and educated person can change the world with one act of courage. The most physically weak person can yield generations of healthy and powerful off-springs. And the oldest person can have a world-changing idea before life ends. Thus, there is no way to describe any person as obsolete despite the ideas of eugenics as introduced Galton [18]. Discarding the idea of eliminating people with arguably inferior genetic traits or even advancing people with arguably positive genetic traits, I will limit the discussion of obsolescence to machines, information systems, social systems, and nonhuman organic systems. Even with these systems, the ways they are measured must capture the systems' total value to include the values in uniqueness. Just because common metrics cannot be established does not mean a unique trait has no value. With due consideration of all the intangibles, the following are some rationales for obsolescence.

3.4.1   Obsolete Because of no More Need

The first reason for declaring a system as obsolete is if the need for the system has completely disappeared. In that case, the system is merely consuming resources and occupying space. Systems that we hope will no longer be needed are weapons of war. Even if we cannot stop all wars, a ban on certain weapons that continue to kill indiscriminately after the war, such as landmines, would make those systems obsolete. The International Campaign to Ban Landmines (ICBL) is working toward this elimination of need [19]. Further, as of 2013, 161 state entities have signed the

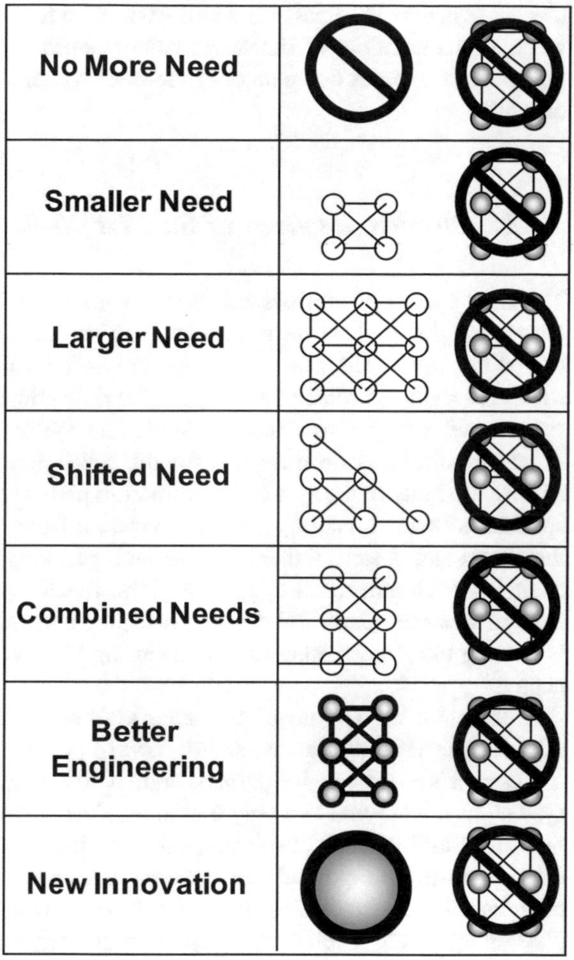

Fig. 3.7 Rationales of
system being obsolete

No More Need		
Smaller Need		
Larger Need		
Shifted Need		
Combined Needs		
Better Engineering		
New Innovation		

Mine Ban Treaty, which was adopted in Oslo, Norway, in 1997. While ending such a damaging need is hard, many needs driven by the preferences of people have evaporated over night.

Hundreds of toys and devices are no longer purchased by people because buyers lost interest. Thousands of patented system concepts were never built because the need for each never emerged. Countless computer applications fall out of the market every year. And small social networks on the World Wide Web form and lose their purpose everyday. Systems that have lost their mission and purpose can linger on for quite some time. In some cases, they remain operational and must be consciously turned off. In other cases, they become dormant or discarded but will remain unless there is disposal. Returning to the example of land mines, these obsolete systems from past wars will have to be found and properly destroyed even when all the countries of the world have stopped using land mines. Thus, we will explore how to break down an obsolete system after we look at what makes systems obsolete.

3.4.2 Obsolete Because of Smaller Need

The need for a system does not have to go away in order for the system to be obsolete. If the need has been reduced, then the existing system might not be a good fit for the need. The need for a system is typically expressed in the form of requirements. Requirements will specify the functions that must be satisfied by the system and the performance standards the system must achieve. Performance standards often include measures for the reliability, accessibility, maintainability, and vulnerability of the system in addition to performance measures associated with specific system functions. Some systems have pre-established performance thresholds and objects. Other systems have performance that evolves and adapts to changing environments. Regardless, if the functional needs become fewer and/or the performance standards become lower, we must ask whether the current system is wasting too many resources or consuming too much time in meeting the smaller need.

If there are ways to acquire another system to meet the smaller need at lower total cost, faster time frames, and/or lower operational complexity and risks, then the current system can be perhaps termed obsolete. In such a case, the current system must be retired in a way that allows the newer more aligned system to take over the core functions. Typically, this transition process will include a period of overlap when both old and new systems are operating. This overlap allows for a coordinated transfer of functions either as a whole with validating tests or in phases/increments to allow for adaptation and adjustments. Phased transfer can be function by function or user group by user group. Function by function transfers tend to focus on integration issues and user group transfers tend to focus on scaling issues. Until this transfer is complete, the old system cannot be declared obsolete. The reality is that a system is not obsolete no matter how well it can perform as long as it is the only system that can respond to requirements. In some cases, the miss alignment between the current system and smaller need is great but there are no replacement options. Then, we must weigh the cost of the smaller need and decide whether to simply eliminate that need altogether.

To figure out ways to align to a smaller need, some obsolete systems can be broken apart and reorganized to be more efficient in meeting requirements. Systems with modular design can eliminate some functional components. Systems with scalable deployment can reduce their presence. For example, a chain store can be considered an obsolete system if its market has declined. If people are not buying specific types of items, these items can be eliminated from the store inventory. If the people are not visiting specific store locations, those branches can be shut down.

Some businesses have managed to completely restructure and rebrand themselves in the wake of a bad market environment. Thus, obsolescence does not always mean that a system has to be destroyed, and some breaking of the system can help in survival.

3.4.3   Obsolete Because of Larger Need

In the opposite direction of system misalignments, the need can grow beyond the functions, capacity, and performance range of the current system. The societal area experiencing the highest expansion of need in recent times is perhaps in information technology systems. If a system performs well at one location, then maybe a thousand locations will want to use the same system. The first instinct in responding to the larger need might be to simply deploy the system one by one at a thousand locations. However, information technology does not have to scale in such an inefficient way. If the thousand locations are all connected to the Internet, then a single system hosted remotely can simultaneously support all the locations provided that it is designed to scale to the number of users and handle multisite access. In designing such a centrally hosted enterprise solution architecture, the physical racks of servers on which the software application is running must have enough capacity and load balancing, the network going from the hosting location to the user locations must have acceptable delays (no greater than X milliseconds latency) and capacity (bandwidth), and a software application plus database that can scale to the enterprise. With advances in cloud computing, the hosting infrastructure has become highly flexible and scalable. However, the software and database must be designed for cloud deployment. Therefore, an existing software with excellent functionality can become obsolete when the locations and number of users it must support expands.

Staying with the software example, the users often want to increase the number of functions based upon usage experience, changing conditions of operations, and better understanding of requirements. If the software is written as a monolithic set of codes, the ability of programmers to modify the codes to add features is often limited by the architecture of the software. To overcome such limitations, modern software applications are sometime designed with modular components integrated together through sets of standard interfaces. This then allows functionality to be added as new modules with less potential for adverse impact on total system operations. Still, regression testing is needed to make sure that all original as well as added functionalities are working well together. Systems without such modularity can, therefore, become obsolete quickly when given more and more functional requirements.

The hardest requirements increase for an existing system to respond to might be in performance. Measures such as reliability, availability, maintainability, and vulnerability are often inherent to the design of the system. New security standards can sometimes be addressed through additional layers of security add-ons. Higher reliability standards can sometimes be addressed through more redundant and backup parts. However, these brute force approaches do not resolve the fact that the design of the system has become obsolete. In fact, extra components added onto an obsolete design can hinder access and increase maintainability burdens. Thus, unless a system has been designed to handle an expansion of performance requirements, it will most likely become obsolete when faced with a larger need.

3.4.4   Obsolete Because of Shifted Need

The need for a system does not have to decrease or increase to create an alignment issue with the current system. The need could simply shift within the reference frame for each requirement. For example, an aircraft must be able to cruise efficiently at a new altitude, a satellite dish must be able to broadcast efficiently in a new frequency, or a train must arrive at a different city. The obsolescence of railroad tracks was common as shipping patterns changed and old locations, such as mining towns, disappeared. The abandoned tracks were then left behind as a reminder of a system structure no longer needed. In the case of a shift, the need for trains continues and new tracks must be laid to new factory locations and new population centers. If the new system structures cannot be built, than the train itself instead of just the tracks might become obsolete.

In complex environments, a slight shift in conditions and systems' inability to adapt to those conditions could create problems. For example, some scientists have suggested that a six-degree change in the earth's temperature could cause dramatic climate conditions, rising ocean levels, and changes in water patterns [20]. Many species of animals could become misaligned with the new environment and be forced into extinction. In a new environment, obsolescence can have a domino effect. The decline of one species can trigger a decline in other species. With less vegetation, for example, there will be less herbivores. With less herbivores as prey, there will be less predators/carnivores. The balance stemming from coupled systems can be extremely precise and sensitive. This means that, if one system shifts its way of operations, the supporting or dependent systems must also shift, or all systems might become obsolete. One type of systems coupling is between a mechanical system and its human or computer operators. In the life cycle of operations, the mechanical system could become less reliable or inaccurate, thus requiring greater operator compensation. If the operator cannot advance to respond to this shift in need, then both systems will be useless for the intended mission.

3.4.5   Obsolete Because of Combined Needs

The need for a system could be evaluated with the needs for other systems, and the result might be the conclusion that all the multiple needs can be combined into a single need with more defining requirements to reduce cost, improve integration, and/or simplify development plus operations. This forces the assessment of whether current systems functionalities can be expanded to address a broader integrated set of requirements. Alternatively, a new system design might be better at satisfying the combined requirements. In this competition, some or all of the current systems might be determined as no longer needed.

The ability to integrate requirements is typically dependent upon the similarity, connectivity, and even commonality of requirements. The first benefit in integration is the elimination of overlapping requirements. Then, similar requirements might fall into natural groups for a system to satisfy, and connective requirements might be easier to address through expanded system functionalities. For example, a ship that delivers one type of cargo might be able to accommodate another type of cargo. An airplane that can serve as a fighter might be able to also serve as bomber. A software application that gathers personnel data might also be able to gather medical data. The historical argument against integrating requirements is that a system that tries to do too much sometimes cannot do anything well. However, the economic benefits of have fewer systems often encourages the combining of needs.

3.4.6   Obsolete Because of Better Engineering

While misalignment with need is a rationale for declaring a system obsolete, a system can be shown to be obsolete through direct competition against other systems. Systems competition does not require a fixed reference frame such as that for defining need. Instead, every functionality and performance range of one system is compared with the functionalities and performance ranges of other systems. Through comparison, which may include testing and simulations, one system might be discovered to have better engineering. Even when two competing systems are quite similar, better engineering might extend performance ranges and broaden functionality. As a result, the better-engineered system can be declared the fittest and worthy of survival. Those systems that lost the competition for fittest can be declared obsolete.

In nature, the competition between species usually leads to direct conflict and death. Thus, obsolescence is often not even an issue. Better engineering is a question for man-made systems. The engineering of systems can focus on the quality of parts, the strength of links, the design of the structure, and the integrity of the boundaries. The comparison of systems can be at any of these levels, or it can focus on the dynamics and interactions of the systems. The easiest comparison to make is when a system is intentionally engineered to improve upon another system. In such a case, the assessment is to see whether the improvement is successful and whether additional weaknesses have been introduced. When systems are built to compete with one another but not conflict with one another, then the comparison is more complex. The end result might be different sets of strengthens and weaknesses for each system that lead to a priority judgment regarding which system is better. In building systems with advanced technologies for critical missions, two prototype development efforts are sometimes initiated to minimize risks and to get the best system. An example of this approach was the development of the Advanced Tactical Fighter by the US Air Force with the YF 22 fighter built by Lockheed, Boeing, and General Dynamics and the YF 23 fighter built by Northrop and McDonnell Douglas [21]. After extensive prototype testing, the Air Force selected the F-22 Raptor fighter to enter into production. At that decision point, the YF 23 fighter prototype effectively became obsolete. Are the down selects of competing systems always perfect? Probably not, but thus is a way in which systems are intentionally eliminated.

3.4.7   Obsolete Because of New Innovation

The last and most dramatic way for systems to become obsolete is the invention of new technical innovations that completely change the way systems work. For example, the invention of the light bulb made gas burning street lamps and oil lamps obsolete. The invention of the automobile made horse and buggy obsolete. The invention of the airplane made balloon-based airships obsolete. The invention of the transistor made vacuum tubes obsolete. The list of inventions and the list of obsolete systems due to innovations goes on and on. In each case, we see orders of magnitude changes in system performance, size, resource consumption, and capabilities. The transformations are at times so great that they change the entire societal systems that the innovations support. In these societal systems, the roles of people can be made obsolete by new systems. For example, companies once had computer punch card typists in the early days of computing. People once swept the streets. Bowling allies once had human pin setters. And milkman once actually delivered milk. The obsolescence of job functions does not mean that certain people have no value. It just means that some people have to change careers.

As with better engineering, innovative systems do not need to use a set reference frame to be compared with current systems. However, the comparison often cannot be by metrics to metrics because the new and old systems are so very different. Instead, the measure is the level of transformation as well as the nature and extent of impact. Some impact of new innovations can be negative. The United States used the newly invented atomic bomb to defeat Japan in World War II. However, are nuclear weapons a positive or negative impact on humanity? It did not end conventional wars, and so other weapons of war did not become obsolete because of nuclear weapons. It did not eliminate US adversaries as countries such as the former Soviet Union, Peoples Republic of China, and other nations developed nuclear weapons. So the only thing that nuclear weapons made obsolete is the military strategy of fire bombing cities by conventional munitions. With all things considered, maybe a nuclear weapon is the innovation that should be declared obsolete. A new system does not necessarily mean that it is better, and innovation does not necessarily mean that it is for good.

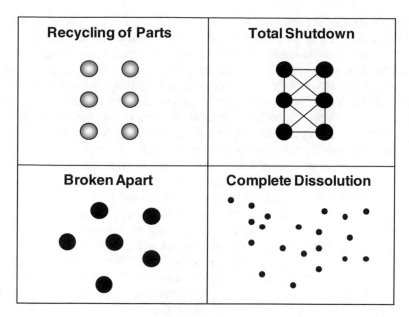

Fig. 3.8 Ways to break down obsolete systems

Assuming that the declaration of obsolescence has been accurately made, the next step is the intentional breaking down of the obsolete systems. Toward this end, there are some obvious strategies, as shown in Fig. 3.8.

The first consideration in breaking down an obsolete system is to determine weather there are any parts of the system that are worth recycling. For example, if there is an obsolete fleet of systems, we might want to retire the fleet in phases over the course of years. In such a case, parts for maintenance might still be required to keep the remainder of the fleet running until it is time for them to also be retired. Alternatively, new replacement systems might have a strategy of intentionally utilizing and integrating with elements of the old systems. A new software system might want to still use the database from the old system to reduce the risk of translation errors. A new organizational system might want to hire employees from the old organization. For any of these reasons, the obsolete systems must be broken down with care so that the parts can be salvaged and recycled. Old system operations can be stopped/turned off. System boundaries, such as the shell of a vehicle, can be removed. System internal associations, such as wires and tubes, can be unplugged, but the parts need to be removed with care. In the case of human organizations where the employees might scatter at the threat of unemployment, retention strategies need to be established to ensure that key employees will accept job offers from the new organization.

If the recycling of parts from the obsolete system is not a requirement, then the obsolete system can be shut down quickly, as long as it is safe and has no adverse impact. Afterwards, the shutdown system, with all parts and associations dead but connected, can be placed in a disposal area. In the deserts of the American West, there are graveyards established for obsolete and shutdown military aircraft from past conflicts. In junkyards and landfills across the world, there are countless mechanical and electronic appliances thrown into piles as the waste of the industrial and information age accumulates. Some complex systems such as nuclear reactors have elaborate shutdown procedures to prevent mistakes such as radiation leaks. In the case of nuclear reactors, the radioactive material from the shutdown reactor must be further contained and stored in a proper manner. For other systems such as software application, the shutdown procedures might need to guard against the theft of valuable data. The remaining codes and data files might still need to be stored in secure locations for future auditing even as the system is removed from the servers. Thus, system disposal often requires great expertise and dedicated resources.

The world does not have enough space for the piles of junk systems, and societies do not have room for the dead remnants of old organizations. Therefore, some obsolete systems must be broken apart for disposal. The breaking process can be similar to the process for recycling parts. However, care in maintaining the integrity of parts is not required. For systems without apparent dangers in breaking down, they can, in some cases, simply be smashed to pieces or crushed. If a system is dangerous, the shutdown system parts must undergo specific disposal procedures. In most cases, it is only a few parts that most undergo careful disposal. Parts with dangerous substances, such as mercury, and parts that can be used to build weapons, such as triggering devices, are clear examples of the need for secure disposal.

Some obsolete systems are so problematic after being shut down that they must be completely dissolved, incinerated, erased, or pulverized to prevent harmful consequences. For example, a facility conducting biological weapons research might become obsolete. To dispose of the facility, the structure and its contents should be decontaminated and sealed. An information system supporting classified military activities might become obsolete. To prevent its secrets from being revealed, all its codes and files should be erased, and all the disks and paper records that are not archived should be shredded or incinerated. The complete dissolution of an obsolete system can also be used to gather raw material from the system. Mechanical systems can be melted down for their metals, electronic systems can be take apart for their rare earth elements such as gallium, and even natural systems such as plants can be chopped up for its medicinal compounds. Natural systems cannot be quite categorized as obsolete. However, they are often broken down because man has decided that their components are more valuable than their continued existence. The most tragic events are animal species hunted to near extinction. Historically, the buffaloes of North American were destroyed to near extinction for their skins, the elephants of Africa were destroyed for their ivory tusks, and the whales of the great oceans were destroyed for their oils. Write or wrong, man has been intentionally breaking down systems of nature for thousands of years, and man will continue to do so at some level to sustain human society.

As I have noted at the beginning of this section, declaring a system as obsolete is not always an easy task. In some cases, this declaration is merely an artificial rationale for the reality that decision-makers found more benefits in the breaking down of the system than in its continued operations. Thus, in the grand scheme of how systems break, this section simply states that some systems breakdown because we intentionally broke them down. If the system resists this intentional shutdown effort, then a conflict situation might emerge. The workers of an organization might protest a plant closing. The operations of a machine might not respond to shutdown procedures. And portions of a distributed information system might linger within the network environment. The effort to break down these resistant systems can be intense and even violent. Systems will not always quietly fade away in the night, and mistakes in the breakdown process can cause major disruptions. This fear has led some decision-makers to allow obsolete systems to linger in the background—to survive longer than their utility.

3.5 Stress: The Silent Destroyer

When it comes to the health of the human body, we know that stress kills. Stress, in fact, is not good for any system because stress by definition is the force that a system must oppose to maintain operations. Stress can be caused by conflict, growth, decay, or any other condition faced by the system. However, stress can just be present in the normal operations of the system without specific external factors. Stress can affect the integrity of the parts, links, and structures. Stress can also cause systems to operate in irregular ways to increase stress and vulnerabilities. Returning to the example of the human body, stress will raise blood pressure, cause weight imbalances, hinder organ functions, and reduce immune functions. Stress might also hinder mental processes and amplify emotional states. Some people might be able to think faster and act with more strength due to the adrenalin released by stress. However, decisions made under stress might not always be perfectly thought out. The ways different people handle stress in everyday life reveal the complex relationship that can exist between systems and stress. Even in man-made systems based on set designs, stress can expose unique flaws in systems. We can statistically measure how a group of common systems will respond to stress. Yet, each system failure will be unique, as two cracks never exactly match.

Since the exact dynamics of system breakdown due to stress is unique to the type of system and even a specific system, the mechanisms of stress are perhaps best described by the patterns in which stress and system dynamics are coupled, as shown in Fig. 3.9. These patterns reveal how stress can overcome a system.

Fig. 3.9 Ways stress breaks
down systems

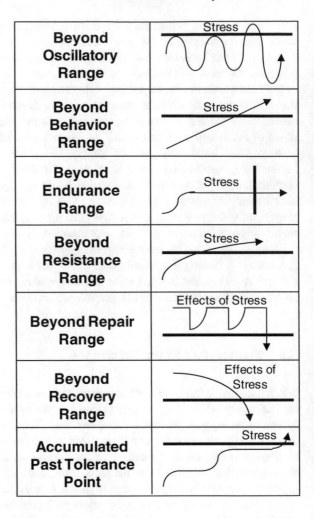

3.5.1 Stress Breakdown from Dynamics Beyond Oscillatory Range

The first pattern in which stress is coupled with system dynamics resides in oscillatory motion. Whether in the physical world, information realm, or societal environment, oscillatory motion in the system involves back and forth changes in the direction of the dynamic vector. This inherently requires force upon the system. For example, a car weaving through traffic will place great friction upon the tires and great force upon the axles and bearings. A computer system undergoing high and low usage loads will have time between each peak to complete actions, but the low periods must be long enough to handle the backlog. And the economic system goes through fluctuations in prices due to the market trying to balance itself between buyers and sellers.

Systems that have oscillatory dynamics typically also have the capacity to handle the forces. The threat of systems breakdown due to stress therefore emerges when the range of oscillation and level of force exceed the normal oscillatory patterns. In the case of a car, a drastic turn at high speed can flip the vehicle, blow tires, and snap structural components. For computer systems, excessive usage loads can cause escalating backlog until the delays become unacceptable and buffers are overloaded. Markets in the economy can go through extreme inflation or depression leading to economic collapse.

Most systems that undergo oscillatory dynamics can also tolerate some degree of dynamic change. However, once the oscillation exceeds a limit, then the stress will break down the system. Understanding this limit is highly beneficial for those who wish to operate systems at dynamic extremes. A good race car driver, for example, will know exactly how hard to push the vehicle in maneuvering around the track. Problem arise when those who are exploiting the limits underestimate the breaking points of systems. In the case of the economy, there is much money to be made in an inflationary market. However, an economic crash will result if inflation is allowed to escalate beyond a point where an oscillatory path can bring about system instability. Once a system starts to break in oscillation, it will break in a system unique way. The pieces of a broken system might be tossed out of the oscillation, or the pieces might come back to the oscillatory path. A breakdown that tosses the system pieces out of the oscillation tends to be sudden and complete. In contrast, a breakdown where some pieces remain on the oscillatory path is sometimes slow, with the breakdown occurring over multiple cycles of exceeding dynamic limits.

3.5.2 Stress Breakdown from Dynamics Beyond Behavior Range

All systems have limits for each type of dynamics that they can sustain. The total of these limits yields a behavioral range for the system. The system might have safeguards to prevent operations from exceeding dynamic limits. However, system control is sometimes given the freedom to push the system's behavioral range. In the case of the human body, for example, the mind is able to push the body into actions of great and unhealthy stress. This freedom has allowed people to overcome great dangers where the risk of bodily stress is outweighed by imminent attacks upon the body. If you are a pilot being chased by a missile, do you not want to fly the plane until the wings fall off to avoid immediate death? If you are running a factory that is about to go bankrupt, do you not want to push production to the breaking point to help alleviate financial losses? Thus, there are rationales for letting systems push beyond behavioral ranges and letting the resulting stress break down systems. Also, some systems are simply so poorly designed that there are no mechanisms to prevent operations from pushing beyond behavioral limits. Other systems might have such basic constraints that they can be circumvented by foolish operators and hijackers.

Once a system is being pushed beyond its behavioral limits, the mechanism of breakdown is unique to the forces/stresses exerted on the system. Typically, the types of behavioral push are speed/rate, capacity, and stretching of capabilities. In the case of speed, it is the force that enables the acceleration and the environmental forces opposing the acceleration that cause stress. Such forces for an airplane are thrust from the engine, form drag caused by changes in airflow around the shape of the aircraft, and friction drag caused by the contact of airflow against the surface of the aircraft. In the case of capacity, it is the force containing the increasing content of a system that causes the stress. In a social rally in which tens of thousands of people have gathered, the forces of leadership and law enforcement are containing the force of chaos within the crowd. However, as the crowd gets bigger and more excited, the stress can cause a collapse of social order and promote violent mob formation. In a system, just about any capability can be stretched past behavior limits. If there is a scale, then that scale can be exceeded with the risk of breakdown.

Sometimes, the behavior limit is exceeded by making a system act in such a way that it is contrary to the intended path of operations. For example, a bunch of people swallowing live goldfish is trying to see how much contrary behavior the body can tolerate. While the body will hopefully give up or throw up before permanent harm is done, there are some risks due to the unexpected stress. Returning to the airplane example, a pilot does not have to fly at mach speeds to break the system. The airplane can be flown to excessive altitudes, turned upside down, or with the wrong aileron configuration to cause damaging stresses. Given the ranges many systems can be made to behave contrary to their intended operations, preventing stress in cases of systems mishandling can be difficult.

3.5.3   Stress Breakdown from Dynamics Beyond Endurance Range

Systems that operate under a state of continuous stress might succumb to the stress in the course of normal operations. In physical systems, material structures can weaken in molecular bonds and deform. In human organizations, workers can reach points of exhaustion and collapse. In information systems, data errors can accumulate during operations until the application cannot handle the errors. For all these and other cases, the system endures the stress to maintain operations. The endurance eventually fails after a period of time, and the system quickly breaks down. Some systems can operate for years and decades under heavy stress, but the reality is that such systems have never fully operated in steady state. Damages accumulate in opposition to stress until they trigger massive breakdown. If a system is exactly the same from one moment to the next under stress, then the point of breakdown is a random event. Endurance then has no meaning, and the association between stress and breakdown becomes difficult to quantify.

System failure under stress is not random, and the failure marks the end of endurance. To understand this failure point, we need to determine when the stress starts acting upon the system and how long the system can endure the stress. This understanding can only come from modeling the specific system or statistically studying the class of systems under stress. The advantage of modeling is that the actual process of breakdown can be brought to light, and ways to hinder the process can be applied. One way to keep systems running past the endurance range is to constantly check for accumulated weaknesses and change parts before or immediately after they fail. Another way to keep the systems running is to harden the parts and associations against stress. Hardening can be by thicker and stronger material, greater than required capacity components, and/or components with barriers against the forces causing stress. For some systems, redundant parts and links can also be added to sustain operations even when there are stress failures. However, redundancy like hardening can add burdens upon the operational system. Even without these efforts, heroic attempts to fix the system could give it added life.

Systems breaking down at the end of endurance can come with or without warning. Systems that are aware of the effects of stress might see signs of fatigue among their parts. This is quite true with human workers. However, fatigue can also go unnoticed, such as hairline cracks in mechanical devices, overheated electronic elements, and longer runtime delays in software. In the breakdown process, the time between detectable signs and rapid breakdown is important in reducing the impact of breakdown. The rate and the mechanism of the final breakdown are also important in determining outcomes. In many cases, it is safer to shut down the system before the final breakdown. Yet, resisting system death, as in the case of the human body, to get the precious additional days of life might be all-important.

3.5.4   Stress Breakdown from Dynamics Beyond Resistance Range

Endurance is the resistance of continuous stress. When the stress upon a system is increasing, the resistance in the system must also be increased to prevent or delay breakdown. The goal of resistance is to get the effects of stress down to a manageable, near steady-state level. However, resistance has its limits, and escalating stress that pushes the system beyond the resistance range will succeed in causing system failure. In the work environment, for example, increasing assignments can cause great mental and physical stress. To overcome the effects of fatigue, people might take stimulants and nutritional supplements to help the body resist. The stress might then be brought to an acceptable level through people working harder and faster. However, if the assignments continue to increase beyond the rate they are being done, the workers will collapse. Resistance is a simple balancing act between forces, but the consequences are severe.

In geopolitics, the history of the Cold War can be studied not as a conflict but as a situation of mutually imposed stress and resistance. The stress was that of perceived strategic and tactical military advantage. The resistance was the development of deterrence weapons and strategic defense weapons. The Soviet Union spent so many resources resisting the threat of the United States that the economic toll contributed to its system collapse. In societal infrastructures, the lack of adequate water supply to support a growing system, such as the city of Las Vegas, is a form of stress. The resistance is then massive construction projects to bring in water from more distant locations. However, if the population and demands continue to grow, the system will break.

The challenge for resistance in complex systems resides in understanding the breaking point because the purpose of resistance is to avoid the breaking point or move back the breaking point. In some cases, the system breaks even as the resistance was about to achieve balanced forces. Resistance can be difficult if the stress on the system is caused by a combination of forces. Then, the resistance might have to focus on the integrated effects and not just the isolated effects of the forces. The failure of resistance, like a dam holding back water, could start with a small crack. Once the stress is able to break through and cause damage, the resistance might not be able to plug that crack. Accurate modeling of stress and resistance forces helps to predict the points and conditions of failure. While failure might be unavoidable if one is only resisting, the impact of failure can perhaps be managed and redirected depending on how one can adjust resistance forces.

3.5.5   Stress Breakdown from Dynamics Beyond Repair Range

The idea of repairing the system to increase endurance was discussed earlier. The process of repair can further be applied to a system prone to sporadic or periodic failures under stress. Stress makes some systems unreliable. In human organizations, for example, some people will have poor physical health and be more vulnerable to stress. Thus, a degree of medical attention must always be available to restore the ill and get the person integrated back into an organization. In mechanical systems, some level of component breakdown will occur during stressful operations even after the initial failures during system infancy. Not all these failures are necessarily due to component flaws, as unique environmental conditions can also cause failures that require repair. Cars that travel across hazardous road conditions might periodically need to have flat tires fixed. Facilities that sporadically get hit by natural disasters might need major repairs over the course of their operational life. The objective of repairs is to keep the system operating at standard levels by resolving the impact of stress-induced midlife part and association failures. The stress might eventually break down the system through decay, but that will be toward the end of life. The repair mission encounters problems when the level of damages and failures exceed the capacity to repair. This can happen if the stress unexpectedly increases during operations or if the effects of stress were not properly accounted for during system design or formation.

A system can go through many cycles of damage and repair until one event exceeds the repair capacity. For example, a car ready to have a flat tire replaced might encounter two or three simultaneous flat tires. An organization ready to handle a few sick workers might have to shut down if half its workforce catches a virus. When stress can fluctuate in such a manner, the decision to expand repair capability must be based on the probability of occurrence, the level of damage, and the cost of preparedness. This cost could be more than financial as extra preparedness can weigh down system operations, delay system start dates, and even introduce other stresses and operational complexities. Alternatively, we can let the damages occur and figure out ways to restore the system after operational disruption. Our next exploration of stress breakdown mechanisms examines the idea that just because a system has been stopped due to stress does not mean that the system has been broken to a point where it cannot return to operations.

3.5.6   Stress Breakdown from Dynamics Beyond Recovery Range

A system can immediately start to decline under stress. The initial decline can just be the system suffering from reduced performance. Then, parts and links break to further reduce performance. Repair attempts can be initiated as discussed, or the system can be allowed to decline to the point where a massive recovery effort will restore system operations. Reasons for allowing system decline include: (1) continuous repairs being too costly or impossible; (2) effects of stress cannot be halted; (3) reduced performance can still satisfy the system's mission; and (4) the system's immediate mission is too important to allow for delayed responses to stress. An example of impractical continuous repair might be the errors that emerge and accumulate when software and databases are aggressively used. A simpler solution could be to do a periodic audit and reboot when the accumulated errors become troublesome. An example of the unstoppable effects of stress is the wearing out of surface material by wind, water, or particular matter. At some point, the damaged system needs to be brought back to the repair shop to have the surface material repaired or replaced. Further, an example of satisfactory performance despite stress is a military unit that can maintain combat effectiveness under heavy casualties. Staying with the military example, commanders have taken planes, ships, tanks, and other vehicles into heavy enemy fire to destroy a critical target at all cost. Normally, the death of systems in combat should be addressed through the conflict mechanisms discussed earlier. However, when damages to the system in combat are ignored and treated as stress, then the mechanisms of stress should perhaps be used to study system breakdown.

The entire strategy of letting damages mount and restoring the system at a recovery point depends on knowing at what point is recovery possible. Even in the case of military sacrifices, the commander must know whether the system will last until the final attack and whether anyone will survive the mission. Understanding the recovery point involves modeling the profile of system decline under stress and having a method of recovery for the profile. If this understanding fails, then stress can easily push the system past the recovery point, and recovery attempts initiated too late will not prevail. Ways that the understanding could fail include unanticipated changes in the stress profile, undetected system vulnerabilities due to stress, and over-projection of recovery capabilities. Any of these failures will lead to system breakdown.

3.5.7 Breakdown from Stresses Accumulated Past Tolerance Point

The last breakdown mechanism based on stress that we will explore is simply the accumulation of many stresses. Individual stresses might not break down a system. However, when stresses can add up in the course of system operations, then they can collectively push the system to the breaking point. There are many stresses in the everyday life of a person, such as bad bosses, home repairs, financial problems, illnesses, accidents, car repairs, etc. If the person cannot resolve these stresses or at least compartmentalize these stresses, then the toll on the body can be tremendous. The same assessment can be applied to any system that operates in a complex environment with many stresses. Once a type of stress hits the system, the first question is, "How long will it last?" The second question is, "What parts in the system does it affect?" When the next stress appears, the questions shift whether the effects overlap the parts and associations impacted by the prior stresses and whether there are additional compounding effects. Compounding effects are more likely when the durations of the stresses also overlap because of the mechanisms causing damages can intertwine. If the same parts and associations are affected by the stresses, then one type of compounding effect is the damages caused by the first type of stress making parts and associations more susceptible to damages caused by follow-on stresses.

The accumulation of stresses creates potential permutations that make determining the breaking point very difficult. The sequence in which the types of stresses hit, the gaps between when the stresses hit, and the intensity of each stress combine to yield hundreds, if not thousands, of possible scenarios with many potential breaking points. Given this inability to project the breaking point, the strategy to prevent system failure should consider decoupling the impact points of stresses through system design, separating the system's vulnerability to stresses through component development, and spacing out the start of stresses through system operations. If necessary, the system should be removed from the environment to which stresses can accumulate. Even slight adjustments in a system's operational profile can help it overcome accumulated stresses.

The many paths of stress reveal that most systems experience some level of stress and that stress can be insidious and harmful. Other mechanisms of system breakdown are events driven and system state driven. Stress, on the other hand, can be continuous, and the effects of stress can be gradual, causing latent patterns of damage until breakdown is occurring in full force. How the human body responds to stress is a major area of medical study [22]. The uniqueness of how each person is physically harmed validates that the effects of stress are closely tied to the characteristics of the system. In the case of the body, physiological response to stress is shaped by genetics, mental attitudes, physical health, past experiences, stress reducing activities, family and friendship structures, and medications/ treatment applied. While heart attacks, strokes, cancers, immune response failures, sleep disorders, substance abuse, and other activities could all be linked to stress, the exact effects of stress on the complex human system still remains a mystery. Thus, stress is a silent killer of systems, and the paths of stress are very important in understanding how systems break.

3.6   Assimilation: The System that Is no More

Systems do not have to be destroyed in order for them to be no longer considered as systems. If a system loses its self-identity by being assimilated into another system, it has in a way been broken. The parts in the system can all be unharmed, and even the associations might endure the assimilation process. However, if the assimilated parts and links can no longer meet the purpose of the original system, then the system has failed. However, that is not always the case, as assimilated parts can help the consuming system to fill new roles. In order for one system to assimilate another, there has to be some level of similarity or compatibility between the systems. Assimilation through similarities is based on the idea that two systems have enough common characteristics in building blocks that the structure of one system can be expanded to consume the other system. Assimilation through compatibility is based on the idea that one system can dominate another system in a way that parts are not harmed but are not necessarily utilized by the dominant system. In both these ways to assimilate, the consuming system must be highly flexible and adaptive. The consuming system must be able to extend its boundaries around the targeted system, exert force on the target, and adjust its links and structures to complete the assimilation.

While engineered systems with highly self-organizing parts can assimilate other systems, the systems best at assimilation are those composed of human elements. Human societal structures have been assimilating one another since the rise of the earliest kingdoms. Looking back into history, the consequences of conflict shifted from the annihilation of the enemy to the capture and enslavement of the enemy at some point in the advancement of social structures. Then, the meaning of enslavement began to advance in a variety of directions. What should societies do with another conquered social system? When such systems had very simple structures and very weak links, the assimilation was about the parts/people. However, after societal substructures such as manufacturing and educational sub-systems gained value, the assimilation had to adjust appropriately. The ways to assimilate complex human societies can be used to help define the mechanisms in which assimilation breaks systems, as shown in Fig. 3.10. These mechanisms are explored below, and the standard for human system assimilation is the loss of freedoms, loss of prosperity, and loss of self-awareness.

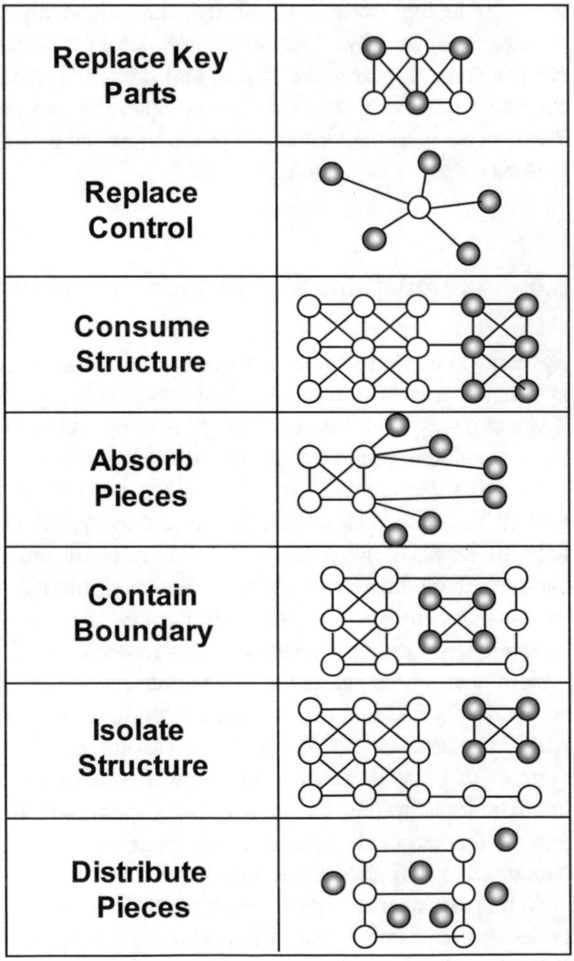

Fig. 3.10 Strategies for assimilating systems

Replace Key Parts	
Replace Control	
Consume Structure	
Absorb Pieces	
Contain Boundary	
Isolate Structure	
Distribute Pieces	

3.6.1   Assimilation by Replacing Key Parts

Once a dominant system has managed to extend its boundary around another system, one assimilation mechanism is to move its own parts into the targeted system without destroying that system's structure. A historical example would be the People's Republic of China solidifying the Chinese border around the region of Tibet in 1951 [23]. Until then, Tibet had existed as multiple kingdoms, each under varying degrees of Imperil Chinese control for hundreds of years. With China's border/system boundary well established by the People's Liberation Army to engulf Tibetan society, the Chinese leadership then supported the migration of Han Chinese (the ethnic majority of China) into the Tibetan region from the late 1970s. This was after the death of many Tibetans during the turmoil of the Cultural Revolution. The Han Chinese immigrants, who initially lived in settlements, built infrastructure, increased economic activities, intermarried with the indigenous population, and formed business relations with Tibetans. Despite Tibetan riots and demonstrations in the 1980s, the Han Chinese number increased to nearly 10 % of the population and became a part of the Tibetan system. From a systems perspective, this insertion of parts into the targeted system enabled successful assimilation. Now, Tibetan independence is nearly an impossible idea. In fact, the debate has switched to how much of the Tibetan culture will survive after another few generations.

The Chinese assimilation of Tibet has been an aggressive and sometimes violent endeavor. Leaving ethical assessments to historians, the systems analysis question is "What is the appropriate rate and level of key parts replacement needed to achieve stable and successful assimilation?" If the Chinese had slowed down in supporting Han migration, would the natural forces of population mobility still assimilate Tibet in a matter of a few more generations? This question connects with the overall assessment of how well small indigenous cultures can survive under globalization. The second part of the question is the number of inserted parts that changes the system. In the case of Han Chinese immigrants, they brought economic wealth and more material goods into the Tibetan system. So, even a small percentage number of immigrants made a significant impact. Further, the Tibetan local government is under Chinese national government control. Therefore, the Tibetan system could not, as a whole, resist the changes occurring. In other areas of the world, majority populations are also expanding into ethnically unique regions but without the tight control that China exerted on Tibet. As population percentages change and tensions increase, the debate is whether such regions are being invaded, assimilated, or evolved. Perhaps the answer lies in the outcome.

3.6.2   Assimilation by Replacing Control

Once a consuming system has engulfed a targeted system, another assimilation process can be by replacing all the control entities in the targeted system. This is very different than compelling the existing control entities to follow the guidance of the dominating/consuming system or just replacing the head of the targeted system. Removing layers of hierarchy from a system can be violent and brutal. Thus, this assimilation approach often needs to follow a conflict between systems and the overwhelming defeat of one side. The advantage of replacing the control structure is that a larger number of parts from the consuming system do not need to be inserted into the targeted system. China could not have done this in Tibet because the opposition by the Tibetan people and unavoidable violence would have devastating effects on China's position in the international community. History, however, has other examples such as the Norman Conquest of England.

Without turning this into a history textbook, Duke William II of Normandy invaded England in 1066 with an army to exert his claim to the throne after the death of the Anglo-Saxon King Edward the Confessor [24]. The duke defeated King Edward's brother-in-law, Harold Godwinson, at the Battle of Hastings and became known as William the Conqueror. Then, the duke seized lands and titles across England and gave them to his followers. This replacement of control was so successful that the duke himself was able to spend much of his time back in France. The control of land by Norman nobility helped to usher in the feudal system, which changed the Anglo-Saxon's way of life from tribal structures to fealty to lord's who ruled from castles. Some historians have argued that this change in English culture was unavoidable because England's Anglo-Saxon culture would have advanced along the same path [25]. Nevertheless, the Anglo-Saxon culture was successfully assimilated and forever altered by the Normans. Even their language of Old English faded into history.

In modern times, with countries managed by either democratically elected leaders or dictators with obedient administrators, the principles of a powerful nobility class seem out of date. Without the willingness of people to yield control, assimilation through a controlling class is difficult to accomplish. In a democracy, the controlling class must pander to the people and work with the attitudes of the people to achieve agendas. In a dictatorship, the people may tolerate control. However, the control will not be able to shift the culture of a conquered people. The most extreme attempt to shape the culture of a people was perhaps the Cultural Revolution in the People's Republic of China from 1966 to 1976 [26]. For ten years, the dictators of the Chinese Communist Party destroyed books, persecuted intellectuals, killed religious leaders, and wiped away cultural artifacts all in an attempt to bring the people/parts of China into a perceived modern system/cultural construct. Despite great material losses, the traditional cultural of China survived because the Chinese people refused to yield. Assimilation through the replacement of the control structure, therefore, requires the inserted structure to be truly integrated with the structure of the rest of the system. The rest of the structure cannot drive the control structure and cannot ignore the control structure.

3.6.3   Assimilation by Consuming Structure

In order for one system structure to fully assimilate another system structure, there cannot be much resistance. Wars and conflicts break down structures in the course of defeating systems. The process of reducing a system's identity without destroying the system's structure requires a gentle cooperative assimilation process. This gentle assimilation process was successfully created during the establishment of the European Union in 1993 [27]. Today, the European Union consists of 28 countries whose national systems are assimilated into a single system of 500 million people. Each country voluntarily submitted itself for entry into the Union. The Union then assimilated a country through the cooperative dissolution of borders, insertion of legal standards, integration into a single market, expansion of commerce, and participation of leaders. Some members further surrendered their national currency and adopted the common European Union currency known as the Euro. In the Union, the structure of each nation is preserved, but the identity of each nation is superceded by the identity of the Union. To the rest of the world, the European Union is a powerful economic system and a formidable political force. The stability of the system sustaining the Union merits debate and more system modeling when considering the government debt crisis in Greece after 2008 global recession [28]. However, from the perspective of the Union being able to consume entire national system structures as a mechanism of assimilation, the Union is a historical example of success.

Two more ancient examples of national structures being gently assimilated are the nations of Mongolia and Manchuria. The Mongolians under Genghis Khan invaded and took control of China as well as much of the rest of Asia to form the Yuan Dynasty in 1271 [29]. With great admiration for the Chinese culture, Genghis Khan's descents sat on the throne of China and allowed the Chinese national structure to extend into the smaller Mongolian national structure. The collapse of the Yuan Dynasty led to a retreat of Mongolians back to their ancestral homeland, However, many Mongolians stayed in southern China and in the lands of the Middle East where they became assimilated. Of the approximately nine million people today who can still identify themselves as ethnically Mongolian, six million people live within the borders of China.

An even more comprehensive assimilation process occurred with the invading Manchurians in 1644 to form the Qing Dynasty [30]. Again, the Chinese culture and system structure was so overwhelming that the Manchurians allowed their national structure to be consumed by the Chinese structure. Though the Manchurians made their brand upon the Chinese culture through four hundred years of rule, the national structure of Manchuria was in the end completely assimilated into China to the point that there is no longer any Manchurian identity left in modern times and extremely few pure blood Manchurians. These ancient examples validate the concept that systems can be broken down by assimilation and that the assimilation process can consume entire structures if the structures willing accept.

3.6.4 Assimilation by Absorbing Pieces

The opposite mechanism to consuming entire structures in assimilation is the breakdown of structures in assimilation. This is slightly different than destroying system structures through wars and conflicts. When a targeted system has already been captured by the consuming system, the breaking down of the structure can be more controlled and without the chaotic destruction of parts. As the associations between the parts of the targeted system are removed, the parts can then be relinked to the consuming system. This absorption of system pieces in the context of empire formation is most effectively achieved among people of similar ethnicity. For example, most empires of the past, such as Rome, China, and Macedonia, started with dynamic leaders that unified a region of feuding tribes/kingdoms with common ethnicity and related culture. Then, because of system similarities, conquered neighbors were not enslaved or killed. Instead, in each case, local system structures were dismantled and an overarching system structure was applied to unify the empire. People from the conquered kingdoms intermarried with the people from the kingdom that initiated the unification, and the unique identities of the once feuding kingdoms faded into history.

The Greeks did the best in preserving their history prior to Philip II of Macedonia and his son, Alexander the Great, forming the Greek Empire around 340 BC and starting the Hellenistic Age [31]. However, Greeks today do not think of themselves as Athenians, Spartans, or Macedonians. Many Chinese today see themselves as the Han people, which extend to the first emperor of China, Qin Shi Huang, who unified the country after the period of warring states [32]. The Han Dynasty, which formed after the death of the first emperor, shaped Chinese culture from 202 BC to AD 220. Although the Chinese dynasties extend back to approximately 2100 BC, no Chinese today traces his or her lineage back to the ancient kingdoms that controlled various portions of the region now recognized as China. Finally, Rome was once merely a city state along with other city states on the Italian peninsula [33]. Though the formation of the Roman Empire is a complex piece of history, and the Italian peninsula again became city states during the Middle Ages, the people of Italy today see themselves as one people and one culture that stem from the conquests and assimilations of Rome.

When a system has been weakened, its pieces can potentially be pulled out and relinked with a more dominant system, even when the weakened system has not been captured by the dominant system. An example of this scenario is the immigration of people from politically oppressed, economically suffering, and situationally unsafe national structures to more prosperous and free countries such as the United States. The system of the United States is in many ways built from the boldest and/or most capable parts from other systems across the world. In some cases, entire social substructures were moved to the United States and transplanted into ethnic neighborhoods. Typically, by the second or third generation, the immigrants in the United States are completely linked into the national system structure. This type of assimilation does not immediately destroy the source countries of the immigrants. However, if the best and brightest of other countries have come to the United States, how could such countries rise above their political chaos, economic limitations, and social issues? In the absorption of system parts, there will always be gains and losses.

3.6.5 Assimilation by Containing Boundary

When a targeted system has no value to the consuming system, the consuming system may opt to contain the boundaries of the targeted system when outright destruction of the targeted system is problematic. Once contained, the targeted system can be allowed to gradually weaken and structurally disintegrate due to a lack of internal resources and access to external goods. The containment might only focus on the system structure. Thus, parts could be allowed to bleed out of the system to quicken the aggressive assimilation. An example of this assimilation and system breakdown mechanism is the containment of the American Indian nations as the United States population expanded westward [34]. The American Indians have already been contained onto ancestral tribal lands through treaties established after the increased arrival of Europeans in the 1600s. To sustain the expansion of the United States after the establishment of the national system, the Indian Removal Act was passed in 1830, which allowed the government to offer land west of the Mississippi River for Indian land east of the river. The most violent removal of Native American people from their indigenous lands was perhaps that of the Cherokee nation. Under President Andrew Jackson, the Cherokees were marched on the Trail of Tears westward where they were contained to reservations.

The treatment of the American Indians was tragic and presents many ethical questions for historians. However, the Native American's commitment to maintaining their traditional way of life made other mechanisms of assimilation difficult and perhaps impossible. Their tribal structures were so tightly integrated that the consuming system cannot insert parts and control entities. Their way of life was so alien that the consuming system cannot connect with their tribal structures. And their people were so different from the rest of the American population that attempts at reeducating Indians for participation in American life was both unpopular and, in many cases, unsuccessful. The American South did not even want Indians as slaves, but instead opted to ship slaves over from Africa. This left three options: (1) destroy the Indian systems through wars and diseases like smallpox, which the Spaniards already did in Mexico; (2) contain the Indian systems in reservations, which the US military did in the mid to late 1800s; or (3) recognize and respect the uniqueness of the Indian systems and do not try to the assimilate. Unfortunately, the story of the American Indian is the story of how systems break. It is a story of how their buffalo food supply was wiped away by American sports man. It is story of decades of poverty. And it is a story of the people surviving and individually triumphing even as the system is lost. American Indians fought in World War II and in other American conflicts. The Navajo language is so unique that it became a WW II communications code that the Japanese never broke. The Apache scouts were so effective that they served in the Navajo War, Mexican Boarder War and WW II.

3.6.6 Assimilation by Isolating Structure

Sometimes, sizable national or ethnic system substructures emerge as opposing entities within greater country systems. These substructures might be brought about by the migration of people, forced relocation of people, or invasion of more dominant people. And these substructures can become so big that containment is either not possible or will not lead to assimilation. Further, the integration of these substructures into the primary system by means of the other assimilation mechanisms discussed could result in changes to the overall system, which are unacceptable by leadership. The alternative strategy to immediate assimilation is then to isolate the substructure until assimilation is either unavoidable or until an acceptable assimilation path can be devised. Examples matching the ways in which the substructures formed can be presented to illustrate delayed assimilation.

For the example of substructures formed by population migration, we can look at the migration of the Jewish people into Europe after the destruction of Jerusalem by the Romans in AD 70 [35]. The diaspora of the Jews led to cultural substructures across the Christian kingdoms of Europe. To block the spread of Jewish influence in European communities, the papacy of the Roman Catholic Church called for the isolation of Jews into ghettos across European cities in 1204. The practice of Jewish isolation would be implemented to varying degrees across European countries until the rise of Nazi Germany. Then, the Nazis initiated a European-wide campaign to kill all the Jews as the most extreme and horrifying approach to prevent Jewish influence on European national systems. The defeat of the Nazis allowed the strength of the Jewish people to contribute to the identity of Europe and enabled the restoration of the nation of Israel.

For the example of substructures formed by the forced relocation of people, we can look at Americans of African descent whose ancestors were forcibly brought to the New World as slaves. After the liberation of slaves brought about by the US Civil War in 1864, the African-American population of the American South was isolated by the policy of segregation [36]. There were black churches, schools, stores, and even hospitals. The South was simply not willing to let African Americans assimilate into the dominant white population. The policy of segregation survived until 1968 when the US Supreme Court declared it as unconstitutional. Through the Civil Rights Movement of the 1960s, minorities and women gained the right to participate in the US system as equals and as integral members. Yet, American society, even today, is far from being integrated both in the North and the South. Neighborhoods often remain predominantly black or white, and the boundaries have many times shifted over the decades with whites moving to the suburbs and younger whites moving back into the cities through gentrification. As some Americans of African descent gained more wealth through better education, the assimilation mechanism is slowly advancing. Now, instances of isolation are perhaps based more on income differences than racial differences. The greatest indicator that the American system is finally ready to accept its black population could be the election of its first black/mixed race president in 2008. The long journey to this point reveals that assimilation is not just about the abolition of isolation boundaries. It also requires the participation of both sides. To some whites in the South, their system and way of life must appear to have totally broken down. Perhaps it is they who must be assimilated into the new American structure.

Finally, for substructures formed from indigenous people by more dominant invading people, we can look at the isolation of the majority black population by white South Africans during the period of Apartheid. South Africa has a complex history involving Dutch, English, and other European settlers along with indigenous black people and slaves from Indonesia. In 1950, soon after the region broke away from British control, the whites of South Africa officially adopted the policy of Apartheid through the Population Registration Act, which classified and segregated people by race [37]. The dominant white population enjoyed a very high standard of living at the expense of the indigenous black population, which constituted 80 % of South Africa. The relationship between the whites and blacks were nothing like that of the Normans and Anglo-Saxons in England because the social structure of the blacks were completely isolated to prevent assimilation. Under great international pressure, the white-led National Party of South Africa began dismantling the Apartheid structure in the 1990s, and the black led African National Congress took power in 1994 with the election of Nelson Mandela as president. The assimilation that has been occurring since 1994 has strained the South African structure. As the majority black population tries to redistribute the wealth of the former structure created by the white South Africans, unemployment rates have risen, whites confronting poverty have increased, and civil unrest has emerged. The white population of South Africa had years to plan for this integration, but the results are showing that assimilation is not easy.

3.6.7 Assimilation by Distributing Pieces

The last mechanism of assimilation that we will explore is a corollary of assimilation by absorbing pieces. Sometimes, it is easy for a targeted system to be broken apart but difficult for its pieces to be absorbed by the consuming system. If the pieces, which are parts with or without link fragments, can be brought to a level of harmlessness, then they can be simply left among the structure of the consuming system. In Europe, the Romani people that migrated over from northern India over a thousand years ago quickly broke a part into wandering bands commonly known as Gypsies [38]. However, the nomadic life of the Gypsies and their unique culture made absorption into primary European cultures quite difficult. Yet the Gypsies did not try to bring down governments, cause social unrest, and/or disrupt the main social system in any way. So most of the European countries viewed them as the wandering poor and allowed them to wander. They faced persecution in some countries, but the great violence levied upon the Gypsies was by Nazi Germany, which viewed them as racially inferior. The Gypsies later migrated to the United States and Brazil, and these countries now have the largest Gypsy population.

Spain attempted to forcibly integrate the Gypsies in the 1600s. The Habsburg Monarchy attempted to forcibly integrate the Gypsies in the 1700s. Norway attempted to forcibly integrate the Gypsies in the 1800s. All these attempts failed, even though the Gypsies have traditionally been quite willing to adopt regional languages and dominant religions. Thus, the only level of assimilation that can be achieved is to let the Gypsies wander and integrate on their terms. Throughout these thousand years, the Gypsies have never tried to organize their people, which could be as many as 12 million worldwide, and the Gypsies have never tried to create a new national system as the Jews successfully did in Israel. Thus, they are perhaps the best example of harmless pieces distributed but not integrated across a system structure.

Assimilation as a method for breaking down human systems can be seen throughout history. However, the mechanisms for assimilation vary greatly in their outcomes, particularly the level of harm rendered on system parts—people who have lost their group identity. Some mechanisms can be fast, as demonstrated in the formation of the European Union. Other mechanism can take centuries, as in the case of Manchurians disappearing into the population of the Han Chinese people. Although historical examples can be used to demonstrate assimilation mechanisms, it is important to note that each event of assimilation is unique because the systems and interactions involved are very complex. The uniqueness makes it challenging to extend historical data to project outcomes of future system assimilation attempts. In fact, some historical events were probably never planned to be forms of assimilation. Some assimilation events occurred instead as a collateral result of system dynamics. For example, as the Manchurian emperors sat on the dragon throne of China, did they consider that the entirety of the Manchurian race would be assimilated through intermarriage in just four hundred years? Assimilation can follow conflict, but assimilation is not conflict. Therefore, the modeling of assimilation should be different. Small changes between systems might have no meaning in conflicts. However, many small changes over a hundred years can still crush a system. We must ponder at what point has our study of people moved from systems science to the social sciences? At what point are models and projections impractical? At that point does assimilation become either a benchmark for history or a strategy for the future?

3.7 Flaws: The Poorly Formed System

We have taken a long journey across many system breakdown mechanisms stemming from obvious methods of how systems break. As with most efforts to divide reality into definable and digestible chunks, there may be other ways to organize our understanding. We could have studied breakdown across all known systems as a function of time and organized our sections by systems that fail in minutes, days, months, and years. We could have studied breakdown based on the level of damage experienced by parts and structures. Alternatively, we could have studied systems breakdown across a class of systems. We can define system classes based on structural characteristics, size or magnitude, functionality, and behavioral characteristics. These are all viable approaches that can unearth hidden patterns and reveal deeper understanding of how systems break. I have chosen to divide systems breakdown based on methods largely because the other ways to organize our understanding of how systems break have so many variations in possible functional groups. Hard to believe, but the group hierarchy of system classes and event types can be larger and more complex than the methods and mechanisms hierarchy shown. This is because the world is filled with so many different systems. To complete our methods and mechanisms hierarchy, the last method for how systems break that I will propose is that systems break because they are inherently flawed.

The method caused by system flaws does not refer to what shows up as a part of decay or what are broken by stress. Instead, it is the system never working correctly from the beginning, and it is the system eventually crashing because of failing operations. In the case of mechanical systems, these flaws are typically discovered during prototype testing and corrected through design modifications. Then, some level of early systems breakdown as discussed in the section on decay is tolerated. In software systems, these flaws are often discovered across the life cycle of system testing, deployment, and use. Once discovered, they can often be fixed by adding patches and replacing code modules. At times, undetected flaws in mechanical systems have led to catastrophic events and major system recalls. Also at times, software systems have yielded major vulnerabilities, such as back doors in system security, which have led to substantial information and financial losses before being repaired. Despite these dramatic dangers, which we hear about in the news, flawed system designs are relatively easy to understand particularly after systems break down. If the computer model for a system was inaccurate at the beginning, forensic data from after the first set of system failures will generally allow for refinements and corrective solutions.

In contrast, understanding flaws in the formation of natural systems and in systems that mimic natural behaviors is a more interesting and challenging endeavor. Systems with self-reacting and self-adaptive parts can yield highly complex structures and behaviors. Evolution Theory is based on the concept that natural systems will have flaws in their continuous attempts at adapting to the environment, and the way nature handles these flaws is the wiping out of weaker systems through competition. Because of the thousands of years required for us to study whether "survival of the fittest" is truly a valid theory, the use of human organizational systems to study breakdown due to flaws is perhaps more practical. Human beings in organizations are highly reactive and adaptive. Therefore, changes that occur on the order of days/months/years can sometimes mimic changes in natural systems.

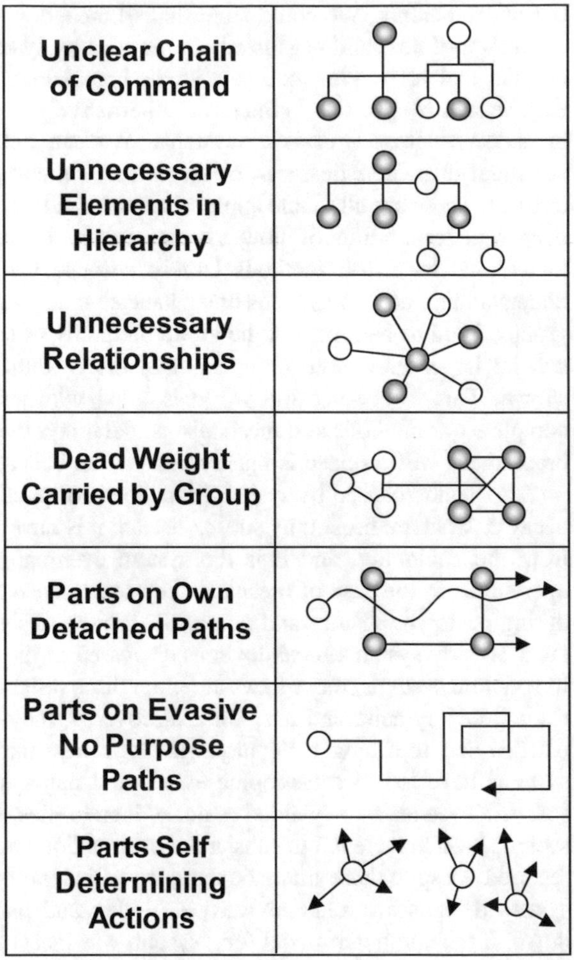

Fig. 3.11 Flaws in the
behavior of people in
organizational systems

Unclear Chain of Command	
Unnecessary Elements in Hierarchy	
Unnecessary Relationships	
Dead Weight Carried by Group	
Parts on Own Detached Paths	
Parts on Evasive No Purpose Paths	
Parts Self Determining Actions	

Fig. 3.11 Flaws in the behavior of people in organizational systems

Using human organizational systems to study how systems break due to flaws yields a set of mechanisms connected with adaptive behavior, as shown in Fig. 3.11. These flaws are all shaped by individual personnel within an organization seeking to manipulate their positions and responsibilities in the organization for personal gains. The misalignment between personal agendas and the mission of the system then creates flaws. Obvious personal agendas include: (1) avoiding of work; (2) pursuing authority and power; (3) increasing financial gains; (4) protecting job security; (5) competing with colleagues; and (6) hiding incompetence as well as mistakes. These agendas are mere motivations. What creates flaws in the system are the mechanisms used to fulfill these agendas. Once a system is formed with one, some, or all these mechanisms in effect, then system breakdown is nearly assured. There is no need for the system to decay over time, and no need for stress to be applied.

3.7.1 Breakdown from Unclear Chain of Command

What should be the most obvious flaw in an organizational system is an unclear chain of command. Conflicting or confusing chains of command can emerge when people are pursuing authority or when matrix organizations are established to allow for relationship changes in determining authority. Unfortunately, in these activities, a great deal of subterfuge could exist to hide the flaws in the system and give the pretense of a clear command and control structure for the system.

In the pursuit of authority, two people or groups vying for command could end up dividing the allegiances within the organization. Some political systems allow for a separation of authorities along with the requirement for collaboration. When the separation is unclear, and when the process for collaboration breaks down, the system can easily enter into gridlock and failure. Even when the system forces one side to officially have command authority, the opposing side could marginally comply with commands and undermine the instruments of control. Competing activities can be hidden from system operations, and these hidden activities can cause dramatic decreases in system performance. In the Middle East and elsewhere in the world, country boundaries have been established by colonial powers to cut across traditional ethnic regions, thus creating countries with competing ethnic populations [39]. Regardless of whether this was done intentionally or unintentionally, an argument can be made that these systems were formed to break down. All the conflicts that have stemmed from such organizations seem to give some credence to this consideration.

In matrix organizations, authority is organized along horizontal groups, such as by regions, and vertical groups, such as by functional areas. Depending on the nature of each new endeavor, either a horizontal or a vertical leader could take control. For example, if a company wants to launch a product across the world, then the executive in charge of the product might gain the highest authority. That executive would then work with company executives in charge of sales in each region. Alternatively, if a company wants to expand into a new region, then the executive in charge of that region might gain the highest authority. That executive would then work with company executives in charge of specific products to bring those products into a region. This process of shifting authority could work in principle. However, flaws in the system will exist when there are situations that encourage competition for credit or a race to place blame. In the case of launching a company product, the executive in charge of the largest regional market might contest the authority of the productive executive, and the product executive might blame a lack of cooperation on the regional executive. In the case of regional expansion, the product executive of the most successful product might contest the authority of the regional executive, and the regional executive might blame the lack of market growth on the productive executive. Once the blaming starts, the system is on a slippery path to breaking down due to flaws.

3.7.2 Breakdown from Unnecessary Elements in Hierarchy

In the command and control of a large organization, authority is often delegated along a hierarchical management structure where each subordinate level is concerned about understanding the commands from higher levels and translating this understanding into more detailed commands for lower levels. This hierarchy permits control down to every specific action taken by people within the organization but does not require the top commander/leader to oversee thousands of actions. The existence of a command and control hierarchy does encourage exploitation by people who either crave status within the organization or want to find a hiding place through falsely justified responsibilities. The results of such exploitation could yield flaws in the system that will force the system to break down.

The potential flaw of people seeking status and power within a hierarchy is that such people might try to inappropriately insert themselves into the hierarchy or to unnecessarily make sure that they are not at the lowest level. An inappropriate position within a hierarchy is one in which a passing of commands to and from the person is not required to translate strategic direction to specific actions. Levels in the hierarchy exist to decompose commands by groups and subgroups. For example, the president of a global company could command the company to increase sales by 10 %. The regional vice presidents would then develop strategies for local managers who would then determine tasks for employees. Now suppose a local manager wants more status and inserts him or herself above other managers. This creates a pass-through and a potential barrier between the vice president and the other managers. At best, such a flaw in the system leads to delays and inaccuracies for passing commands. More likely, however, the rogue manager will manipulate the commands received from the vice president to control other managers and advance personal agenda. As there is no proper role for the rogue manager, manipulation of commands is often the only way for him or her to justify the fake position. One person behaving in this opportunistic way will probably not bring down the system. However, the system will collapse if this behavior is profitable and spreads among the self-centered and ambitious.

When people are fighting not to be at the lowest level of a hierarchy, they might be tempted to create artificial layers of subordinates below them. A person who can perform a task might find a way to justify why the task should be performed by several people. If those subordinates think the same way and figure out how to further delegate, then an unnecessary pyramid begins to form. Just a few extra layers in a hierarchy can add a great deal of burden upon a system. People in hierarchies with well-defined connectivity of responsibilities cannot create many artificial layers. However, some hierarchies in growing systems can have more and more unnecessary layers established in the process of growth. In a massive system about to collapse from excess management weight, one might find that, for every logical level of management, there are two or more layers of authority. These excess layers consume resources, slow down the reaction time of the system, and resist change. However, the solution cannot always be by flattening the management structure through brute force. Instead, natural system layers should be discovered and ways to select and converge managers to those layers must be formulated. In many cases, the system will run out of resources and collapse from the unnecessary weight.

So far, I have talked about unnecessary elements due to personal ambition. However, people can also make themselves unnecessary by avoiding responsibilities. Then, their positions in hierarchies become hiding places. For example, a manager might have real responsibilities to oversee subordinates but chooses instead to be a mere figurehead. A person might find a way to define a position not for status but for the legitimization of having no responsibilities. Having no responsibilities might not harm the system unless the system has to support these positions. Support includes drawing finances, consuming resources, tying up communications, and complicating processes. If people all start to evade responsibilities and hide within hierarchies, the system will fail. Large companies with complex management structures fight this force every day and often end up being defeated by smaller innovative companies with a take-responsibility culture.

3.7.3 Breakdown from Unnecessary Relationships

We have explored flaws based on confusion in the management structure and unnecessary burdens in the command and control hierarchy. Unfortunately, flaws in an organizational system can start with any person seeking to change his or her relationships within the organization. Most organizations require people with specific roles and responsibilities to communicate and coordinate with other people to form an integrated process. In organizations that do not have to respond to changing operating conditions, the relationships between people can be clearly constrained through standard operating procedures. For example, person A on the assembly line will give a completed item for person B to install within every X minutes. If the speed of the assembly needs to adapt to situational demands, then the procedures must more flexibly allow for management guidance. If person A must at times change the construction of an item and give to person C, then even more flexibility must be allowed.

The risk of flexibility in systems with people/parts that can adapt is the self-generation of unnecessary relationships. We see this phenomenon with the availability of computer-based document and meeting management tools. Documents that once only required a few people to review and approve are suddenly sent to over a dozen people. Meetings that once were held between a few people suddenly have many additional virtual attendees. This growth of relationships serves people's self-interest by distributing accountability and diluting blame if mistakes are made. At the same time, more people can share in the recognition when efforts are successful. Therefore, workers in many cases all like expanded relationships.

I do believe that groups can achieve many great things and that computer tools have facilitated the exchange of ideas as well as the coordination of actions. Even more eyes on activities can help to identify critical mistakes before major impact. In the context of how systems break, these benefits make it more difficult for us to identify unnecessary relationships in the system. At what point does the number of involved people cause those who should taken responsibility to be careless? At what point in the size of collaboration groups do people start to build outputs based on consensus and the least common denominator? These are serious flaws in a system, and the effectiveness of organizations in the computer and network age is still a grand experiment. Organizations with established global processes have taken great advantage of expanded relationships. However, very little data has been collected on the flaws presented.

3.7.4   Breakdown from Dead Weight Carried by Group

I have already noted that the command and control structure of a human organization can be burdened by unnecessary elements. A more observable flaw in the system is having entire sections as dead weight. These sections could have at one time been useful within the organization. If so, the changing processes of the organization and/or advances in technology have caused them to be left without roles. Business forces within companies will often drive out employees whose roles of have disappeared. However, powerful unions can delay business forces and adjustments of salaries to align with market conditions. In governmental organizations and institutions, the force of bureaucracy can also delay the elimination of dead weight. Two artificial rationales for keeping people whose roles have disappeared within an organization are: (1) the mass of people is too heavy to cast out; and (2) the class of people is too integrated to cast out.

For example, skilled machine operators in a factory might band together to oppose being laid off when robot assembly systems are installed. Individually, the workers can be eliminated as robots are transitioned in. However, as a group, the workers can threaten to disrupt the transition process unless a bargain is reached. The bargain might allow the workers to be retrained for other functions in the organization, or the bargain might create a tragic systemic flaw where the rest of the organization must carry a set of workers with no defined responsibilities. The organization will continue to carry this weight until it gradually breaks down because, somehow, it has determined that a total separation is too unstable.

As another example, a group of senior professionals in an organization might have capabilities that are outdated compared with those of younger graduates. However, these senior professionals are in positions where they can cause serious disruptions if they were to be removed. They could refuse to pass organizational knowledge to their replacements. They could give trade secrets to competitors. They could alter processes to cause failures after they leave. Thus, despite that organization's need for better capabilities, the threat and fear of disruption hinder the elimination of dead weight. Instead, organizations might consume excess resources to hire younger graduates as additional workforce. These additional employees will, however, face internal opposition from those senior professionals who have outdated capabilities. In some cases, the number of seniors in organizations will grow over time as many resist retirement and layoff attempts. Organizations have collapsed under this critical flaw in the system. And so, dead weight in the system is a mechanism for how systems break.

3.7.5 Breakdown from Parts on Own Detached Paths

Some systems are composed of parts tightly linked together, and some systems allow parties to be highly independent when interacting with one another. But again, systems are, by definition, parts working together. When a part does not work with other parts in the system and follow its own path, that flawed behavior will interfere with the activities of other parts and create a gap in the system process if that part had critical responsibilities. The ways in which a part can become detached from the associations and activities of the system include: (1) inability to keep up with the changes in the system; (2) inability to truly satisfy its intended role; (3) individual agendas requiring detachment; (4) external forces triggering detachment; and (5) rejection by other parts in the system. These ways and their mechanisms of breakdown can also be illustrated with human organizations.

In military structures entering the battlefield, individual soldiers can often become detached from their units both geographically and in planned actions. This detachment is typically defined by a loss of communications, uncertainty about physical location and orientation, and confusion regarding the intent of the commander. Yet, the soldiers may feel compelled to act in the heat of battle. This need to act and need to fend for self-survival can lead soldiers to fire accidentally on friendly troops, give away situational details to the enemy, interfere with the maneuver of own troops, and be captured by the enemy to be used as leverage. Thus, military systems that cannot maintain control of all its units and parts during the chaos of battle will have a critical flaw. Systems that might lose communications should instill soldiers with preset strategies and instructions for maneuver and attack.

In organizations with members who are not qualified, the members might try to create a set of activities just so that they will look like they are a part of the system. In the meantime, work that should be done by the members is not done. System performance will then drop if the management does not have a good way to filter effective work from pointless busy work.

A person in an organization might want to detach from the rest of the system to satisfy a personal agenda. If the person feels betrayed but still needs what the organization provides, then he or she might continue on in a detached manner. If the person has a personal life issue such as death in the family or divorce, then he or she, too, might continue on in a detached manner. Detached people can still perform organizational functions. At a reduced level of concentration and energy, however, the work quality and error rates of these people must be questioned.

Some parts in a system might be forced out by external forces. When the associations in an organization become too complex or dynamic, people who cannot keep up might start to fake their associations and secretly detach. The types of associations that are easy to fake are paper work, which can be filled out quickly and poorly, requests that can be granted without proper consideration, and observations that are not really made. Any of these fake associations could result in performance errors that break down the system.

Finally, a person might be rejected by other people in his or her organization, even though his or her services are of critical need. When a system is formed of people, all its parts may not act in the best interest of the system. Many organizations officially ban acts of discrimination based on race, gender, age, and orientation. However, isolation from the primary group of organizational coworkers can still occur. The isolated person might remain remarkably dedicated to the organization, but his or her effectiveness will be reduced without the ability to collaborate. This critical flaw in the system has been recognized in many modern societies. Yet, discrimination was still historically condoned in many societies, such as in the United States and South Africa. This changed with events such as the Civil Rights Movement of the 1960s and Abolishment of Apartheid in 1994.

3.7.6 Breakdown from Parts on Evasive no Purpose Path

The corollary of parts being separated from a system is parts evading integration with a system. In an organization, a member's role, capabilities, and/or knowledge may be recognized by others in the organization. However, attempts at working with the member and leveraging his or her skills have failed because the member does not want to be a part of the organization. This can happen with members who have been forced into service or with members who are forced to stay in an organization. The most apparent example of forced service is perhaps captured enemy scientists. Unless they have been successfully motivated, these intellectuals can produce all manner of falsely credible work that deceptively pushes the organization off track. In the same manner, many slaves will do as much as possible to evade their intended roles in the organization. While the minimal performance of a single slave might be acceptable, the commitment by all slaves to evade responsibilities will break down the organization. Historically, the attempts to enslave some groups, such as the American Indians, have failed because they would rather evade and resist to the point of annihilation than being enslaved.

The most apparent example of members forced to stay in an organization is perhaps people who signed commitment letters or contracts but later regretted the decision. These people could be corporate executives during a business buyout, entertainers for a set number of engagements, and soldiers deployed to combat zones. In all these cases, the regretful people will try to minimally satisfy their obligations but evade truly meaningful integration with the organization. Minimal satisfaction might be acceptable in some cases but could be a problem if the organization is undergoing high dynamic stresses. For example, companies in highly competitive environments do not want executives who are not focused on the competition. Entertainment industries with very selective audiences do not want entertainers who are not focused on courting the paying customers. And military forces in high threat areas do not want soldiers who are focused only on self-survival and getting home. These parts will create critical flaws in the system, and the evasive path of these parts does not even have to be completely without purpose. The system might break simply because the paths of critical parts are not aligned to the right purpose.

3.7.7 Breakdown from Parts Self-Determining Actions

The final mechanism of system breakdown that I will propose is the insidious flaw of parts in a system determining actions without synchronizing with the rest of the system. In complex organizations, we want people to have initiative and proactively identify problems, find solutions, and act in ways that improve the overall performance and situation of the organization. Therefore, some level of autonomy must be granted to organization members depending on their roles and responsibilities in the organization. The risk of flaws in system operations resides in the misunderstanding of people regarding how they should treat the autonomy and the inability of the system to monitor plus guard against autonomous behaviors that exceed acceptable ranges.

When people are given autonomy in an organization, they must have a clear sense of what observations, assessments, decisions, and actions they are accountable for, the ranges of actions they can take, and the objectives they are trying to achieve. These constraints might not help them achieve the best outcomes for the organization, but they will prevent extremely unacceptable activities short of people intentionally acting in opposition to the organization. If these constraints are not established, people with autonomy might overreact to situations, act against events beyond their sphere of official responsibility, be apathetic against pressing issues, and act on issues and opportunities already being addressed by other members of the organization. When situations are tense, such as two armies pointing guns at one another, a member/soldier overreacting to perceptions and acting prematurely has started wars. In banking companies, a financial manager exceeding his or her delegated authority in monetary transactions could cause regulatory violations that lead to heavy fines and perhaps personal jail time. In government organizations, civil servants will often be apathetic about being proactive because their jobs are only threatened by severe mistakes. If one does not act, then one cannot be blamed for actions. In sales-driven companies with heavy personal bonuses, multiple employees could go after the same richest opportunities. Internal conflicts have emerged out of greed that puts the company in a bad light. These short examples show that autonomy or freedom requires constraints. Systems with unconstrained freedom for their parts will have flaws that lead to breakdown.

A system with constraints upon freedom must have ways to monitor and enforce those constraints. In democratic societies where freedom is the highest, monitoring of people's behaviors includes police patrols, cameras in urban areas, financial audits of personal taxes and companies transactions, filters in social media networks, and screening through worker tests. Then, to discourage behaviors that will harm other people and the societal system in general, there are varying types of punishments, which include fines, community service, imprisonment, and executions. The vast number of people in the United States who are in prison, nearly 1 % of the population, raises the question of whether current techniques are successful at preventing the system from breaking down [40]. Further, the need to deter the most serious infractions on social constraints in the US has sustained the practice of capital punishment. The idea that a system must kill its parts to maintain order reflects the seriousness and consequences of freedom. Yet despite these attempts at constraining behavior, the flaws in human society in the form of major crimes have endured. So human society as a system might yet one day collapse on its own not due to wars, growth, or even decay but due to flaws.

References

1. Korten DC (1995) When corporations rule the world. Kumarian Press, West Hartford
2. Wallerstein I (1974) The modern world-system I: capitalist agriculture and the origins of the European world-economy in the sixteenth century. Academic Press, New York
3. Lenski GE (1966) Power and privilege: a theory of social stratification. McGraw-Hill, New York
4. Harrison P (2006) Post-structuralist theories. In: Aitken S, Valentine G (eds) Approaches to human geography. Sage, London, pp 122–135
5. Hassan I (1987) The postmodern turn, essays in postmodern theory and culture. Ohio University Press, Oxford
6. Gaddis JL (2006) The Cold War: a new history. Penguin Books, New York
7. Chen BX (2014) Q. and A. on Heartbleed: a flaw missed by the masses. New York Times, April 9
8. Britten Austin P (2000) 1812: Napoleon's invasion of Russia. Greenhill Books, South Yorkshire
9. Fried J (2012) Who really drove the economy into the ditch?. Algora Publishing, New York
10. Doyle W (1990) The Oxford history of the French Revolution, 3rd edn. Oxford University Press, Oxford
11. Campbell Bartoletti S (2005) Black potatoes: the story of the Great Irish Famine, 1845–1850. Houghton Mifflin Harcourt, New York
12. Baker S (2007) Ancient Rome: the rise and fall of an empire. BBC Books, London
13. Morgan D (2007) The mongols, 2nd edn. Wiley-Blackwell, Hoboken
14. Wingfield B (2010) The end of the Great Recession? Hardly. Forbes, New York Sept 20
15. Gladwell M (2002) The tipping point, how little things can make a big difference. Little, Brown and Company, New York
16. Ebeling CE (1997) An introduction to reliability and maintainability engineering. McGraw-Hill Companies Inc, Boston

17. Callaway E (2010) Telomerase reverses ageing process. Nature Online, 28 November
18. Galton F (1904) Eugenics: its definition, scope, and aims. Am J Sociol 10(1):82
19. Faulkner F (2007) Moral entrepreneurs and the campaign to ban landmines. Rodopi B.V., New York
20. Lynas M (2008) Six degrees: our future on a hotter planet. National Geographics, Washington DC
21. Goodall JC (1992) The Lockheed YF-22 and Northrop YF-23 advanced tactical fighters. America's stealth fighters and bombers, B-2, F-117, YF-22, and YF-23. Motorbooks International Publishing, St. Paul
22. Everly GS Jr, Lating JM (2012) A clinical guide to the treatment of the human stress response, 3rd edn. Springer, New York
23. Hao Y (2000) Tibetan population in China: myths and facts re-examined. Asian Ethn 1(1): 11–36
24. Crouch D (2007) The Normans: the history of a dynasty. Hambledon and London, London
25. Richardson HG, Sayles GO (1974) The governance of medieval England from the conquest to Magna Carta. Edinburgh University Press, Edinburgh
26. Kraus RC (2012) The Cultural Revolution: a very short introduction. Oxford University Press, Oxford
27. Staab A (2013) The European Union explained: institutions, actors, global impact, 3rd edn. Indiana University Press, Bloomington
28. Lynn M (2010) Bust: Greece, the Euro and the sovereign debt crisis. Bloomberg Press, New York
29. Brook T (2013) The troubled empire: China in the Yuan and Ming Dynasties. Belknap Press, Cambridge
30. Rowe WT (2009) The great Qing. Harvard University Press, Cambridge
31. Martin TR, Blackwell CW (2012) Alexander the Great: the story of an ancient life. Cambridge University Press, Cambridge
32. Clements J (2007) The first emperor of China. The History Press, Stroud
33. Forsythe G (2006) A critical history of early Rome: from prehistory to the first Punic war. University of California Press, Oakland
34. Nichols RL (2004) American Indians in US history (The civilization of the American Indian series), 10th edn. University of Oklahoma Press, Norman
35. Goodman M (2008) Rome and Jerusalem: the clash of ancient civilizations. Vintage, London
36. Price S (2006) When will I get in?: Segregation and civil rights (American history through primary sources). Raintree, Eustis
37. Clark NL, Worger WH (2011) South Africa: the rise and fall of apartheid, 2nd edn. Routledge, New York
38. Taylor B (2014) Another darkness, another dawn: a history of Gypsies, Roma and Travellers. Reaktion Books, Eustis
39. Kamrava M (2013) The modern Middle East: a political history since the First World War, 3rd edn. University of California Press, Oakland, CA
40. Mahapatra L (2014) Incarcerated in America: why are so many people in US prisons? International Business Times, 19 March

Chapter 4
The Systems Analyst

Abstract The concluding chapter provides a summary of the challenges and great opportunities in systems analysis. It humbly describes the magnitude of the discipline and recognizes that there are other ways to organize a conceptual framework for studying systems. Finally, the burden of seeing beyond the range of current data and the constraints of current schools of thought is passed to the reader.

My hope is that systems analysis will one day be taught as a primary subject like math or physics. In this manner, professionals across many fields will gain new perspectives and new tools for confronting their challenges. The world will still need dedicated systems researchers just like the world still needs mathematicians and physicists. However, the popularization of this mysterious discipline will help overcome the first barrier in systems studies. To elaborate, I deeply believe that the current contribution of systems studies is limited by the knowledge gap between systems analysts who do not have an in-depth understanding of specific subjects and subject matter experts who do not have enough understanding of systems analysis to see its benefits. This barrier is being taken down on a case-by-case level through systems researchers who are willing to spend time studying other disciplines and through experts in select fields who are willing to take the risk of trying systems-analysis methodologies. Cross-disciplinary teams have formed, and the broader contributions of systems analysis are being realized. However, some fields and some experts are still committed to disciplinary stovepipes, traditional methodologies, and rigorous credentials for research participation.

The stove piping of disciplines is the second barrier for systems researchers who want to solve problems in various fields. In research communities, where advancements have been slow and in research communities where there are contending schools of thought, great institutions have formed to protect the established paths of research and career advancement. I will not name fields and topic matters out of respect. However, it is not difficult to identify select fields where those who control the research funding within government and foundations subtly require proposals to be from recognizable research groups and principal investigators to have studied and received degrees from specific institutions and professors. These

© Springer International Publishing Switzerland 2017
C.H. Ren, *How Systems Form and How Systems Break*, Studies in Systems,
Decision and Control 72, DOI 10.1007/978-3-319-44030-9_4

standards typically carryover to the peer-review processes for journals where specific groups of peers will only recommend manuscripts from researchers and research activities they recognize. Even in a blind review process, it is not difficult to recognize the affiliation of the authors by checking the selected references and by tracing the topic matter.

This protective mentality in fields with contending schools of thought have progressed to such extremes that dedicated journals have emerged to cater to specific research groups, in essence declaring to the scientific community that the journal only cares about scientific advances if they agree with the journal's theories and philosophies. Those in the field who can advance the current path by 10 % are applauded. Those who might transform the field and create new paths are blocked. Why would some experts want to block breakthroughs that will over turn the foundational theories? I can propose two reasons. The breakthroughs have been so slow in coming that the existing experts have based their careers on the current theories and research paths instead of on the broad potentials of their field. And the promises of additional opportunities after the overturning of current theories are so uncertain that existing experts cannot see how they will fit into the new path. Ironically, I can delve deeper and use systems analysis to explain how some research communities might resist theories and discoveries gained through systems analysis.

This book, however, is not written to criticize those who reject the systems research journey but to celebrate the potential of those who are willing to charge forward. If we lack knowledge about the subjects to which we want to conduct systems analysis, the World Wide Web is filled with information to help us learn and conduct research. If we are blocked from presenting systems research findings, new interdisciplinary journals are emerging to challenge closed communities. I truly believe that we are heading toward a golden age of learning where great discoveries can be made not just by the academic elite but also by the inquisitive students.

Therefore, the last barrier I want to discuss is that of our own hesitation. This hesitation stems from how many of us are taught to pursue the possible. In the case of systems research, we look for problems where there are available data and identifiable solution approaches. Sometimes, we back away from asking the hard question of what is missing. Data is critical in research but also a logical trap because every type of system in the world at one time started without being measured. If we stare only at the available data long enough, we might never see the hidden systems that have not been measured. We might forget that there are more things in this world that we do not understand than things that we do understand. So ask what is not making sense, what has not been explained, and what might be true. It is difficult to do research without adequate data, but it is not impossible to conduct systems analysis. Even when there is not enough data to defend a theory, systems analysis can yield conceptual models that introduce possibilities. Possibilities can lead to new ways to discover and measure potential systems, and possibilities can lead to new reference frames for measuring known systems.

I end with the affirmation that I am humbled by the world of systems and the declaration that what I write by no means capture all that is systems analysis.

The methodologies of systems analysis are ever advancing. The number and quality of system models are changing the world of knowledge. And the capabilities of systems researchers far exceed my ability to understand. I spoke of challenges and great opportunities, but systems thinking is sometimes just a new way of see everything in our daily lives. When we are trying to understand situations and when we are making decisions, systems analysis can help show us all the interrelationships and all the potential consequences. For aspiring students, the decision on how far to push the boundaries of systems analysis is yours to make. There is much to be done in fields that already rely heavily upon systems research. There are teams to join, problems to be solved, and gains to be made. I wish you all success and many years of exciting analysis.

Printed in the United States
By Bookmasters